Intergenerational Challenges and Climate Justice

Climate change poses questions of intergenerational justice, but some of its features make it difficult to determine whether we have obligations of climate justice to future generations. This book offers a novel argument, justifying the present generation's obligations to future people.

Livia Ester Luzzatto shows that we have intergenerational obligations because many of our actions are based on presuppositions about future people. When agents engage in such intergenerational actions, they also acquire an obligation to recognise those future people as agents within their principles of justice and with that a duty to respect their agency and autonomy. *Intergenerational Challenges and Climate Justice* also offers a way to circumvent the problems of non-identity and non-existence. Its approach overcomes the intergenerational challenges of climate change by meeting three necessary criteria: providing ways to cope with uncertainty, dealing with the complexity of climate change, and including future people for their own sake. The author meets these criteria by adopting an action-centred methodology that grounds our obligations of justice on the presuppositions of activity. This robust framework can be used to justify increased climate action and the greater inclusion of future-oriented policies in current decision-making.

This book will be of great interest to academics and students concerned with the issues of climate and intergenerational justice.

Livia Ester Luzzatto completed her PhD in politics from the University of Reading, UK. Her research focuses on the intersection between climate change, ethics, and business challenges.

Routledge Studies in Environmental Justice

This series is theoretically and geographically broad in scope, seeking to explore the emerging debates, controversies and practical solutions within Environmental Justice, from around the globe. It offers cutting-edge perspectives at both a local and global scale, engaging with topics such as climate justice, water governance, air pollution, waste management, environmental crime, and the various intersections of the field with related disciplines.

The *Routledge Studies in Environmental Justice* series welcomes submissions that combine strong academic theory with practical applications, and as such is relevant to a global readership of students, researchers, policy-makers, practitioners and activists.

Climate Change Justice and Global Resource Commons
Local and Global Postcolonial Political Ecologies
Shangrila Joshi

Diversity and Inclusion in Environmentalism
Edited by Karen Bell

Environmental Justice in the Anthropocene
From (Un)Just Presents to Just Futures
Edited by Stacia Ryder, Kathryn Powlen, Melinda Laituri, Stephanie A. Malin, Joshua Sbicca and Dimitris Stevis

John Rawls and Environmental Justice
Implementing a Sustainable and Socially Just Future
John Töns

Intergenerational Challenges and Climate Justice
Setting the Scope of Our Obligations
Livia Ester Luzzatto

For more information about this series, please visit: www.routledge.com/Routledge-Studies-in-Environmental-Justice/book-series/EJS

Intergenerational Challenges and Climate Justice

Setting the Scope of Our Obligations

Livia Ester Luzzatto

Routledge
Taylor & Francis Group

LONDON AND NEW YORK

earthscan
from Routledge

First published 2022
by Routledge
4 Park Square, Milton Park, Abingdon, Oxon OX14 4RN

and by Routledge
605 Third Avenue, New York, NY 10158

Routledge is an imprint of the Taylor & Francis Group, an informa business

British Library Cataloguing-in-Publication Data
A catalogue record for this book is available from the British Library

Library of Congress Cataloging-in-Publication Data
A catalog record for this book has been requested

ISBN: 978-1-032-19377-9 (hbk)
ISBN: 978-1-032-19379-3 (pbk)
ISBN: 978-1-003-25890-2 (ebk)

DOI: 10.4324/9781003258902

Typeset in Times New Roman
by Apex Covantage, LLC

To Enrico and Federica

Contents

1 Introduction   1

Climate change and future generations 1
Justice and beneficence 3
    The non-identity problem 4
    The non-existence challenge 5
    Climate change as a matter of intergenerational justice 6
    Discounting for time 10
Fixing the scope of climate justice 12
A framework for intergenerational climate justice 16

2 Dealing with uncertainty   24

Justice and future climate changes 24
Risk, uncertainty, and ignorance in the context of climate
    change 26
    Theoretical overview 26
    Empirical overview 27
The consequentialist approach to scope 28
    Overinclusiveness 30
    Underinclusiveness 32
An alternative approach 35
    Foreseeability 36
    Autonomy conditions 38
    The duty to respect other as agents 39
Understanding foreseeability 42
Consequentialism revisited 46
Conclusion 48

3 Climate change as a complex problem of justice   53

The problem of complexity 53
The road to a solution 57
An action-centred solution 59

Justice, uncertainty, and complexity 59
Collective agents, intentions, and assumptions 59
Discerning the presuppositions behind organised
  collective actions 61
Intergenerational actions 62
Intergenerational obligations 68
Scrutinising the presuppositions behind collective actions 70
Overcoming the complexity challenge 73
Conclusion 74

4 Including future people for their own sake   80
The moral value of future persons 80
Ends and means 84
A game analogy 89
The importance to us of the afterlife in its own right 91
  How much do we really care about the afterlife? 91
  Care and the justification of justice 94
Conclusion 95

5 An account of the scope of climate justice   98
Taking stock 98
A preliminary account of scope 99
  Dealing with uncertainty 100
  Dealing with complexity 101
  Including future people for their own sake 102
Scheffler's afterlife conjecture and intergenerational
  actions 102
  The afterlife conjecture 102
  Three categories of intergenerational actions 103
  The afterlife conjecture and the scope of justice 105
Which future generations? 106
  Near future generations 107
  Remote future generations 107
Intergenerational cooperation and the scope of justice 109
  Weaker intergenerational actions 110
  Stronger intergenerational actions 111
  A nuanced approach to intergenerational actions 111
An account of the scope of climate justice 113
  The account, revisited 113
  The non-identity problem and non-existence challenge 114
Conclusion 119

6   Changing perspective   122

The problem 122
The argument 124
  The framework 124
  The account 126
Changing our perspective 127
  Theoretical issues 127
  Rethinking our climate policies 128
  *Tackling climate change as an intergenerational
    problem 130*
  Intergenerational actions and democratic institutions 132
Where to go from here 134

Index   137

1   Introduction

Climate change and future generations

Climate change has been termed "the largest, most pervasive threat to the natural environment and human societies the world has ever experienced" (United Nations Environment Programme 2015). Starting with the Industrial Revolution, anthropogenic emissions of greenhouse gases have impacted the earth's climate and increased ocean and surface temperatures (IPCC 2014: 2–4, 2021: 5–8). These changes have already harmed many present people. Yet the effects on future people are likely to be even more severe: the climate will continue to change, largely because of the warming locked in by past and ongoing greenhouse gas emissions. This means surface temperatures will increase further, and extreme weather events will likely become even more frequent, severely threatening the health, safety, and even livelihoods of future people (IPCC 2014: 10, 65).

The intergenerational dimension of climate change also has many distinctive and complicating features which a moral and political response to it must address. These are factors like the temporal dispersion of its causes and effects, the magnitude and severity of the future effects of present actions, and our unavoidable uncertainties about the future impacts of climate change. With these challenges, the intergenerational dimension of climate change tests the preparedness of our political institutions and theories to recognise and respond to the intergenerational problems they have allowed to develop. So far, however, the results have been disappointing. Existing institutions have proven unable to account for the future effects of our actions and devise a response that can ward off the most severe impacts of climate change. While the global community has recognised the need to limit warming well below 2°C (United Nations Climate Change 2015, 2021), current mitigation policies and pledges are not nearly ambitious enough to keep us within that target (Climate Analytics and New Climate Institute 2021). To make matters worse, even staying within a 2°C limit is not enough to avoid severe impacts on future natural and human environments (IPCC 2018).

Given the predicted future effects of continued climate change, it is tempting to understand its intergenerational impacts as a problem of justice. Roughly, one could say that our actions – our overconsumption of greenhouse gas emissions as a "scarce good," as well as our failure to adequately adapt to climate change – are

DOI: 10.4324/9781003258902-1

going to cause significant harm to future people and affect their ability to meet some of their most fundamental needs. Because our actions harm future people in ways that limit their ability to fulfil their most fundamental necessities, they violate our duties of justice to them. This interpretation is intuitively appealing, but, as will become clear throughout this book, the claim that future generations are the rightful subjects of climate justice requires further justification before we can use it to argue for urgently needed climate action.

This book provides such a justification by formulating an account of the scope of our obligations which clearly shows that, and why, we have obligations of climate justice to future people. The scope of justice defines "the range of persons who have claims upon and responsibilities to each other arising from considerations of justice" (Abizadeh 2007: 323). Fixing the scope of climate justice is a crucial component of any defence of intergenerational climate justice: in order to have a useful discussion about intergenerational climate justice, we first need to know to what extent, and for what reasons, we have obligations of climate justice to future generations.

Correctly determining who owes something to whom under principles of climate justice is a question of basic moral concern. Morality requires, for example, that we be able to justify why some people rather than others ought to bear the costs of climate change. It also requires us not to arbitrarily disadvantage certain groups in a theory of climate justice, for instance because – like future generations – they are unable to make their voices heard. An account of scope allows us to satisfy these basic moral requirements by providing us with a sound set of justifiable principles that determine who is owed duties of climate justice by whom.

Fixing the scope of climate justice can also have significant practical implications by affecting which climate actions we can most forcefully advocate for in practice, as well as promoting our understanding of climate change as an intrinsically intergenerational problem. Arguments in favour of including future people within the scope of climate justice can, for instance, be used to advocate for starker climate action aimed at safeguarding the interests of future generations. In turn, a sufficiently thorough assessment of the intergenerational scope of our duties can help us fully understand the ways in which, through the character of our underlying actions, climate change both creates and relies on morally relevant connections to future people. A better understanding of the intergenerational nature of climate change can potentially increase acceptance of and compliance with our intergenerational moral obligations.

In this book I am going to argue for a strictly deontological account of the scope of climate justice. I will show that the scope of our obligations is determined by the character of our actions and not solely by their consequences. More specifically, I will argue that at least one set of our obligations to future people stems from the fact that we are engaged in intergenerational actions. These actions, as I will argue, are actions which presuppose the agency of future people and therefore ground obligations of justice to them. This action-centred account is a particularly good fit for climate justice because it can successfully deal with three issues that are central to intergenerational climate change, namely the uncertainty about future climate changes, the moral complexity of climate change, and the question of how, if at all, future people ought to be included within the scope of climate justice. This book is structured around these three central issues: in the following chapters, I am going to argue that each of these problems gives rise to a separate criterion which a successful account of the scope of climate justice must meet. I will argue that an account of scope must be able to deal with uncertainty, respond to the complexity of climate change as a moral problem, and include future people as ends in themselves if it is to accurately respond to the problem background of climate change and its intergenerational challenge.

Justice and beneficence

Throughout this book, I will use the term "climate justice" to mean that domain of justice which covers agents' obligations and entitlements in relation to anthropogenic climate change. I understand principles of climate justice as protecting all entitlements of justice that are vulnerable to anthropogenic climate change. These can be very broad: they can range from food, water, and shelter to issues of social and economic justice related to forced migration, loss or damage of property, and loss of livelihood caused by climate change. In turn, climate justice includes both principles to avoid or mitigate the negative impacts of climate change on people's entitlements and principles to determine how the costs, or burdens, of doing so ought to be shared between agents in ways that respect what they are owed.[1] *Intergenerational climate justice*, accordingly, refers to the obligations of climate justice that agents of one generation have towards persons belonging to a different generation. In the following, I will use it to refer more specifically to what present agents owe future people.

This book addresses two sets of theorists. On one side are those who are already committed to seeing future generations as subjects of climate justice (see, for example, Caney 2008, 2009b; Gardiner 2011, 2017). While I will argue in favour of intergenerational obligations of climate justice and thus in favour of these writers' overall conclusions, I believe they have often failed to convincingly elaborate the reasons we have for extending climate justice to future generations. I will say more about some of their work later. For now, what matters is that this shortcoming has left these accounts vulnerable to objections against intergenerational climate justice and to being sidelined in policy discussions in favour of cost-based approaches.[2] By giving a detailed analysis of both whether and why we owe climate justice to future generations, this book offers some necessary and additional support to these theorists' conclusions.

On the other side are those who deny that future generations can be the subjects of climate justice and argue instead that our intergenerational duties to them are better conceptualised in terms of beneficence (see, for example, Beckermann and Pasek 2001; Broome 1992, 2012). This has led some of these theorists to advocate for very high discount rates to be applied to future climate burdens, and consequently for slower and less decisive, but presently cheaper climate action (for example, Nordhaus 2008). Opponents of intergenerational climate justice have often relied on one of two main objections: the non-identity problem and the non-existence challenge. This book aims to show that, despite these objections, our duties to future people to address the impacts of climate change fall under the scope of justice and are thus, among other things, not subject to being discounted for time. In order to address these scholars' arguments against intergenerational climate justice, it is important that we first understand their reasons for objecting.

The non-identity problem

John Broome's is one of the earliest, most prominent and extensive accounts of climate ethics.[3] It is also an important example of an argument against intergenerational climate justice based on the non-identity problem.[4] While Broome argues for wide-ranging duties of beneficence to future people, he denies that large collective agents, such as public entities, may owe duties of justice to future generations.[5] He denies the possibility of such intergenerational obligations of justice because he takes non-identity considerations seriously. The non-identity problem affects the way in which we may rightfully think of our relations and obligations to future generations and is brought to bear whenever, in framing our obligations to future generations, we claim that particular future persons are harmed – on the traditional, counterfactual understanding of harm as being made worse off – by our actions.

The non-identity considerations at the heart of the problem go back to the work of Derek Parfit (1984). They point to the fact that many present actions, including virtually all large-scale policies, influence the lives and relationships of present people. In doing so, they affect when and with whom children are conceived, causing certain future persons to be born rather than others. The resulting problem points to two facts. On the one hand, some of our actions will cause future people to live lives that are worth living, but which only offer a very low quality of life; on the other hand, those actions are what cause these people to exist. Future people thus live unavoidably flawed existences: there is no way for *them* to live higher-quality lives, only for a different set of people to do so. The upshot is that future people cannot claim to have been harmed – or to have been made worse off – by an action when it is the case that they would not have existed without it. If we are concerned about the non-identity problem, we either need an alternative notion of harm or a different way to explain what is morally bad, if anything at all, about causing people to live unavoidably flawed existences.

To illustrate the problem in practice, consider the following simplified climate change scenarios. In scenario A, our lives are largely shaped by an economy based on fossil fuels. We regularly take long-haul flights, perhaps work in an emission-heavy branch of industry, and rely on private cars for transport. In this scenario, we meet certain partners with whom we conceive children at set times, thus creating a specific set of future persons. Though their lives are worth living, these persons and their descendants live flawed existences because they suffer the impacts of climate change – a problem caused by the same activities which, by impacting our surroundings, made our lives unfold in ways that caused us to conceive them.

In an alternative scenario B, we succeed in making the lifestyle changes needed to limit our greenhouse gas emissions and ward off most of the future impacts of climate change. However, these lifestyle changes lead us to conceive at different times, possibly with different people, hence bringing a different set of future persons into existence. Although scenario B offers a much higher quality of life to future persons, the same individuals who will live under scenario A could never have existed under these improved circumstances.

This is where the non-identity problem bites. It holds that an action which does not make a person worse off, and which that person would not regret as she owes her existence to it, cannot be wrong because of how it affects that person's interests or rights, as it does not in fact do so (Parfit 1984). In turn, theories of justice that ground duties on violations of subjects' interests or rights cannot give rise to justice-based claims on the part of these future persons. For our climate example, this means that climate changing activities cannot give rise to intergenerational claims of justice based solely on the negative impact climate change is going to have on future individuals' interests or rights. Duties of beneficence, conversely, are widely thought not to be vulnerable to the non-identity objection as they are impersonal duties to do good rather than special duties tied to a specific person's rights or interests (Broome 2012: 66–67). Supporters of the non-identity problem, including Broome, have thus tended to support beneficence over justice as a way to frame collective obligations of climate justice to future people.

Philosophers have long grappled with the non-identity problem and have proposed several arguments to establish intergenerational duties of justice in spite of it (Cohen 2009; Kumar 2003; Page 2008; Rivera-Lopez 2009; Velleman 2008; Woodward 1986). It is important to know that these arguments exist, but I am not going to discuss them here. For the time being, I will assume that among these there are successful ways to overcome the objection and that it therefore does not pose an obstacle to this project. I am later going to argue that my account of scope offers one more way in which the non-identity problem may be circumvented. Roughly, this is because the obligations of climate justice I identify are rooted in the presuppositions we make about future people as people regardless of their particular identities. These obligations are owed to future people as such rather than tied to particular individuals, their rights, or their interests. I will elaborate on these arguments in Chapter 5. For the moment, it is important to note that by arguing in favour of obligations of climate justice to future people this book addresses an important objection against the very possibility of intergenerational justice.

The non-existence challenge

The second objection which opponents have voiced is known as the non-existence challenge. It holds that we cannot have obligations of justice to non-existing future people since justice requires its recipients to possess some form of interests or rights, and this, per definition, can only be the case for currently existing persons. What this challenge calls our attention to is that future people cannot have rights or interests right now, which poses a problem for those who believe their having interests or rights is a necessary condition for establishing present obligations of justice to them (see Beckerman and Pasek 2001; De George 1979; Herstein 2009; Macklin 1981).

This objection is significant but nevertheless simpler to overcome than the non-identity problem. Proponents are clearly right to point out that non-existing persons cannot possess anything at this point in time, when our obligations of justice to them are determined. The most promising way to overcome this challenge is to concede this first premise but show that intergenerational justice does not require the interests it protects to exist at the same time as the respective obligations, in other words that present obligations do not require present interests. For this kind of response, what matters is that we can plausibly assume future people are going to have interests once they come into existence and that these interests are the type of interests that give rise to obligations of justice (Elliot 1989; Meyer 2016).

In order to establish intergenerational duties of justice that can withstand the non-existence objection, an account of scope must therefore show that our obligations to future people are based on something other than their present interests. The account I am going to formulate in this book offers a way in which this requirement can be met. On this account, agents' obligations are triggered by the presuppositions on which their own actions are based and are justified by a general obligation to respect others' agency and autonomy. As with the non-identity problem, I want to set the non-existence challenge and my response to it aside for the time being and will return to it in Chapter 5. For the moment, what is important is that we are aware this challenge exists and acknowledge that a successful account of the scope of climate justice should not require the protected interests to exist at the same time as the respective obligations.

Climate change as a matter of intergenerational justice

Whether we frame our response to climate change in terms of justice or beneficence makes a considerable difference to the force and standing of our resulting obligations. Throughout this book I will rely only on a very thin definition of justice, according to which principles of justice provide, for those who fall within their scope, a set of enforceable and impartial rules that mediate between individuals' conflicting claims to certain material or non-material goods (Miller 2017). Principles of justice apply in conditions of moderate scarcity, which obtain whenever there is enough of the relevant goods to be distributed between claimants so that all have at least a chance that their basic needs be met yet not enough to fully satisfy everyone's desires (see also Barry 1978; Hume 1987 [1777]; Wenar 2017).

Duties of justice cover a person's most fundamental entitlements and give rise to particularly stringent and enforceable obligations (Miller 2017; Valentini 2013). Yet justice covers only part of our moral duties. The distinction between duties of justice and duties of beneficence, which represent another important set of moral duties, is significant. Duties of justice most often take lexical priority over competing obligations, including those of beneficence. They are also commonly seen as exempt from being discounted for time, whereas duties of beneficence need not be.[6] An argument based on justice rather than beneficence is therefore going to be more forceful in arguing for more aggressive climate action and promoting the corresponding changes to our political theories and institutions.

There are important reasons why we should think of climate change as requiring a stronger justice-based response. One is that the impacts of anthropogenic climate change threaten a set of interests that are fundamental to human lives. According to current scientific consensus, changes in our climate are highly likely to have one or more of the following effects: reduced access to potable water, food, shelter, and other basic human needs; displacement and loss of livelihoods; loss of cultural goods and traditional heritage as a consequence of displacement; as well as endangering people's lives, health, and possibility to attain an adequate standard of living (IPCC 2018; United Nations Environment Programme 2015).

These impacts alone are sometimes deemed sufficient to trigger duties of justice (Caney 2008; Nussbaum 2003; Sen 2004). On *recipient-based views*, duties of justice are activated whenever a person's fundamental interests are not fulfilled (Valentini 2013). Fundamental interests cover basic human needs that are universally shared and, on all plausible accounts, include things like nutrition, livelihood, health, and life, all of which are endangered by the impacts of climate change. The mere fact that climate change is going to have such detrimental impacts on quality of life thus in itself provides one good reason to frame the problem as one of justice and not just beneficence.

There are, however, additional reasons why we should think of climate change as a matter of justice that relate to its anthropogenic nature. On the *autonomy view* of justice proposed by Laura Valentini, in order to be under a duty of justice it is not enough that another persons' fundamental needs be threatened. It needs to be something an agent has done or is doing which is threatening another's interests, and that agent needs to be able to foresee and control the effects of that action. On Valentini's account, "each person has a right not to be deprived of the social conditions to lead an autonomous life," and, in turn, "[a]n agent is under a duty of justice if and only if she has the ability to refrain from undermining the necessary conditions for others to lead autonomous lives ('autonomy conditions')" (Valentini 2013: 496). Valentini's own autonomy conditions include mental and physical health, available resources, and social goods like liberties, wealth, opportunities, and services (Valentini 2013: 497). On this view, agents bear duties of justice only if they are able to refrain from undermining others' autonomy conditions. For this to be the case, two conditions need to be met: agents must be able to anticipate the effects of their actions on others' autonomy conditions (*foreseeability condition*) and to control these actions (*control condition*) (Valentini 2013: 498).

This view is appealing in the case of climate change as it picks up a central feature of the problem which a recipient-based view does not. Climate change is mostly caused by human activities that are controllable and whose effects, in the form of climate change and its impacts, foreseeably endanger the autonomy conditions of vulnerable present and future persons. There seems to be something particularly objectionable about the fact that these dangerous changes to the climate are caused by human actions. There seems to be, in other words, a morally relevant difference between a purely natural disaster and one whose underlying cause we can pinpoint as human activity, and the autonomy view does better than the recipient view in picking up this difference.

While my arguments in this book do not rely on either view to begin with, they will highlight the importance of the character of our actions in fixing the scope of justice and will in turn support the autonomy over the recipient view. For the moment, however, I only need these brief sketches of each view to make a small but crucial point: whichever view we later adopt, it is important and philosophically warranted to think of climate change as a problem of justice.

This in itself is not a novel argument. While some theorists still rely on the non-identity problem or the non-existence objection to oppose the possibility of intergenerational climate justice, many before me have already taken the opposite stance and framed climate change and its effects on future people as problems of justice. Sketching some of these contributions to intergenerational climate justice will help place my account of scope within the field's broader context. It will also show why this book adds an important piece to the puzzle of climate justice that has so far been missing. That is, I believe existing accounts of climate justice have not said enough about why its scope should include future people. Many theorists who tacitly endorse intergenerational duties of climate justice, or who have argued for specific substantive interpretations of intergenerational climate justice, have not sufficiently justified why future people ought to fall within the scope of our obligations. This shortcoming has left intergenerational climate justice open to objections like the non-identity problem and the non-existence challenge.

One of these accounts of intergenerational climate justice comes from Simon Caney. Caney takes a recipient-based view of justice and has, in a number of publications, defended a human rights approach to intergenerational climate justice, arguing for the need to view the future impacts of climate change as a threat to the human rights of future generations (Caney 2008, 2009a, 2009b, 2014a, 2014b; for a similar approach, see Meyer 2003). His concern with the basic interests of future generations has been echoed by scholars like Clark Wolf (2009). Wolf's theory of intergenerational climate justice is based on a hypothetical social contract that includes future generations. Also taking a recipient-based view, he argues in favour of protecting the basic needs of future generations, which he posits are most threatened by the effects of climate change (Wolf 2009).

On both accounts, the scope of climate justice is assumed to be intergenerational because climate change threatens the basic needs of future people. Caney's account briefly tackles the question of scope directly, but the discussion collapses into the question of whether present people may discount the rights-claims of future people based only on their temporal location (Caney 2008). Caney argues, I believe rightly, that they may not (Caney 2008, 2014b). This is an important conclusion, but it does not exhaust all that needs to be said about the scope of our obligations. To the contrary, we could make a stronger case against discounting if we could give additional reasons, beyond the effects of climate change on future needs, why our obligations to future people fall under the scope of justice.

In his account of climate justice, Stephen Gardiner identifies a number of plausible ways in which intergenerational climate justice differs from more familiar intragenerational problems of justice (Gardiner 2003, 2004, 2006, 2011, 2017). I am later going to argue that some of these differences, especially the complexity of climate change as a moral problem, significantly affect how we can go about determining the scope of climate justice and that an account of scope must be able to deal with the complexity of climate change in order to be successful.

Gardiner hypothesises that the unique moral challenge of climate change lies in its nature as a "perfect moral storm," which combines three distinct and significant problems of justice (Gardiner 2006, 2011). These separate yet deeply connected issues are the global storm, which represents our inability to cooperate globally and across borders with one another; the intergenerational storm, which calls out our failure to adequately take account of future generations; and the theoretical storm, caused by our current lack of both policies and moral theories capable of responding to the global, intergenerational nature of climate change.[7]

The convergence of these three sets of issues has important consequences for our ability to mount an effective ethical response to climate change. For the purpose of intergenerational climate justice, two of Gardiner's conclusions are particularly interesting. One is that the perfect moral storm makes us, present generations, vulnerable to moral corruption. In other words, it undermines our ability to act morally towards future generations. Because we lack access to the relevant moral norms, we are tempted to distort the norms we do have so that they justify action which benefits ourselves over future generations. Moral corruption may, for example, be what is at play whenever economists apply particularly high discount rates to cost-benefit analyses, so as to justify passing on a larger share of the climate burden to future generations (Gardiner 2011: Chapters 7 and 8). The other is that the lack of adequate norms specifically incites intergenerational buck-passing, in other words the deferral of climate action from one generation to the next. In the face of the challenges posed by climate change, for which we lack appropriate moral guidance, we respond with inertia and self-deception, use the lack of norms to justify buck-passing, and avoid addressing the problem head on.

Gardiner's work is particularly helpful in clarifying the complex problem background against which we are trying to achieve climate justice. However, it too, like Caney's account, presupposes obligations to future people without adequately justifying their intergenerational scope. Gardiner glosses over the question of scope with a general assumption that most moral theories accept at least some substantive obligations to future people (Gardiner 2011: 155–156). While this may be sufficient for an analysis of the problem background, it is not enough to show that, and why, present generations ought to take action on climate change and resist their urge to pass the buck to future people. It is also neither enough to show whether, if they do have such obligations, these are obligations of justice (as I will show they are) nor enough to reject objections such as the non-identity problem and the non-existence challenge.

Because the need for intergenerational justice is urgent, the claim that we have obligations of justice to future people to tackle climate change requires a more convincing vindication than what these theorists have proposed. In this book I try to provide such a vindication by developing an account of the scope of climate justice that is accurately tailored to the problem background of climate change. What I have said so far has highlighted that climate change poses a very particular set of problems that any account of climate justice, including an account of its scope, must be able to respond to. I believe these problems can be summarised as a distinct *intergenerational climate challenge*. The intergenerational challenge emphasises that a successful moral response to climate change must be able to account for the temporal dispersion of its causes and effects; the magnitude and severity of the future effects of present actions; people's varying vulnerability to climate change, which depends on their temporal and geographic location as well as socio-economic status; and the inherent and unavoidable uncertainty about future impacts of climate change. An account of the scope of climate justice must be able to meet this intergenerational challenge in order to provide a strong basis for a substantive account of climate justice. In this book I will argue that an account of scope which meets the three necessary criteria – which can account for future uncertainties, can deal with the complexity of climate change, and includes future people for their own sake – will be able to successfully overcome the intergenerational climate challenge and the more fine-grained problems it calls our attention to.

Discounting for time

I have argued that showing that our intergenerational obligations are duties of justice and not beneficence is going to affect whether the corresponding future claims are discounted for time and that it will therefore impact how our obligations are weighed against competing duties. Let me elaborate why this is the case. A claim is positively discounted for time if it is given less weight, in comparison, the further ahead it lies in time. There is an important distinction between discounting the value of resources and discounting the importance we give to people depending on their position in time. The argument for positively discounting the value of resources is based on the assumption that, given interest rates and economic growth, the value of resources grows with time. Present resources are therefore worth more than future resources, and applying a positive discount rate to future resources reflects this difference (Broome 1994; Caney 2008).[8]

We can therefore see why, if we were to frame the impacts of climate change purely in terms of economic costs, future claims are likely to be discounted. Whether we discount people's moral standing, on the other hand, is based on whether we exhibit something called *pure time preference*. If we do, in our weighing of respective claims we give less weight to those of people in the future based solely on their position in time (Caney 2008). Because claims of justice are directly linked to a person's moral standing, it is generally accepted that we should not exhibit a pure preference in the case of justice. Claims of beneficence, on the other hand, are more often framed in terms of welfare or costs and often, though not always, subject to being discounted for time.[9]

Theorists who measure the impacts of climate change in terms of welfare or costs most often work with a consequentialist, maximising moral theory. Their most forceful argument in favour of a positive discount rate for time is thus one based on economic growth and the demandingness of present sacrifices: if morality requires us to maximise welfare or the availability of some future good, then unless we discount future welfare and resources for time, present people will be required to make immense sacrifices whenever these sufficiently increase future value (for example, if their sacrifices benefit more than one generation). Such excessive sacrifices, however, cannot be what morality rightfully requires of present generations (Lomborg 2001; Posner 2004).[10] Such arguments about discounting are especially relevant in practice, as policymakers often rely on economic approaches based on cost-benefit analyses (Stern 2007; Nordhaus 2008; but see Dennig 2018). Whenever these analyses use higher discount rates, they will favour slower, more modest action against climate change.[11]

In this book I argue that we have significant reasons to frame future claims in terms of climate justice rather than costs or welfare. The account of scope I will argue for gives rise to claims based on future people's agency and autonomy. These are closely linked to future people's moral standing and should therefore not be subject to pure time preference. I will also argue that what matters most in determining and fulfilling our obligations to future people is the character of our actions, not their consequences or role in maximising some impartial good. In other words, the aim of this project is to develop an account of the scope of climate justice which meets the intergenerational climate challenge, and my arguments will show that this is a deontological account which takes seriously the equal moral standing of persons as well as the values of autonomy and agency.

The account I will argue for decouples our obligations from the maximisation of welfare or material goods. Instead, I will argue that we have an obligation to reshape our actions in a way that respects the values of autonomy and agency. Because our obligations are primarily linked to agents' fundamental entitlements to agency and autonomy rather than the availability of material goods, the claims of future people will not be liable to being discounted for resource growth. The very concept of justice also allows us to push back against the argument from demandingness. Within a theory of justice, claims to the fulfilment of fundamental interests will be held by members of both future and present generations, which limits how much present people may be required to sacrifice for future people. Principles of justice are not going to require excessive sacrifices from present people as both the duty-bearers' and the claim-holders' entitlements of justice must be respected in the process of redistribution.[12]

Because it affects how we deal with the issue of discounting, whether or not we frame the claims of future people in terms of justice is going to affect these claims' overall weight within a theory of climate ethics. Insofar as these weightings then affect the actual climate policies that are proposed and implemented, framing our obligations to future people as non-discountable claims of justice can make a considerable difference to how aggressively present generations act to prevent even more dangerous climate change. Correctly determining the scope of climate justice is thus an important task for policy practice and for theory.

Fixing the scope of climate justice

While intergenerational climate change poses a complex challenge for theories of justice, there are important reasons to think more carefully about the scope of climate justice and, most importantly, why future people ought to be included in it. The goal of this book is to provide readers with a sound account of the scope of climate justice that shows whether and why future people fall within the scope of our obligations and responds to the most pressing concerns raised by climate change and its intergenerational challenge.

In the following chapters I am going to use an action-centred methodology to show that we have obligations of climate justice to future people. This method grounds our intergenerational obligations on our actions and, more specifically, the presuppositions on which our actions are based. Roughly, I am going to argue that we have obligations to future people because our actions are based on presuppositions about their capabilities and vulnerabilities. Because our actions rely on presuppositions about future generations, if our actions and moral theory are to be coherent, we must also extend the basic requirements of justice to future people.

The purpose of this investigation is not to construct a full account of moral standing but to develop a focused account of the scope of climate justice for a specific set of agents. In other words, I will formulate an account that explains whom agents relevant to climate justice must take account of when engaging in actions relevant to climate justice. Because this approach will be tailored to respond to the particular problem of climate change that we are facing, many of my arguments are going to rely on facts about our actions and the actual circumstances in which they happen, including, for instance, what we do and do not know about climate change, its causes, and its possible effects on the future. Nevertheless, the account I am going to propose will ultimately be based on a general obligation to respect agency and autonomy. It should therefore also be applicable to other domains of justice, provided it is modified where needed to match that domain's relevant actions.

My investigation will focus on the obligations of climate justice that are owed by *organised collective agents* relevant to climate justice. The "we" whose intergenerational obligations I am going to assess in this book thus covers all organised collective agents that are engaged in actions relevant to climate justice. As climate justice protects those entitlements of justice that are vulnerable to anthropogenic climate change, actions relevant to climate justice are all those actions that directly affect the extent to which we are able to respond to anthropogenic climate change.[13] I take the relevant types of actions to be all actions or omissions by collective agents that meet this criterion and are consciously performed and intended by agents. They can also include attitudes, policies, and practices.[14] Some examples would be the adoption of an adaptation plan for a coastal area, a government's decision to build a coal-powered plant, or a corporation's switch to the use of renewable energy.

I will focus on collective agents for two main reasons. One is empirical, in that a large proportion of actions relevant to climate justice are attributable to organised collectives like governments, international organisations, private corporations, or civil society. Showing that, when engaging in these actions, agents are bound by obligations of justice to future generations, and that they must reshape their actions in a way that meets the requirements of climate justice, can thus potentially underpin a significant shift towards more sustainable policies. The second reason is philosophical and relates to Broome's arguments and the non-identity problem. Broome argues that the non-identity problem applies most forcefully to large-scale collective agents like governments as their actions have the greatest potential to effect changes in who is born. Individual actions have a much smaller reach and are less likely to have such large-scale effects on society (Broome 2012: 64–68). The more interesting reply to the non-identity problem will thus be one that zooms in on collective agents rather than individuals. I have already laid out why I believe it is important to frame climate change in terms of justice rather than beneficence, and the fact that Broome singles out collective agents to challenge this claim makes it especially important to focus my counterarguments on collective agents too.

The problem-focused and action-centred approach to fixing the scope of justice I am going to take draws on the work of Onora O'Neill (1996; see also 2000, 2001). O'Neill argues that we should move away from trying to define self-contained categories of ethical concern and instead develop a justifiable practical procedure to fix the scope of our obligations for each problem as it arises. This is enough for practical purposes and does not rely either on unvindicated moral assumptions or, importantly, on an idealised notion of agents (O'Neill 1996: 97). O'Neill's rejection of idealisations is a key component of her account of scope. Her objection to the use of idealised concepts relies on the sharp distinction she draws between idealising and abstracting. Idealising "ascribes predicates . . . that are false in the case in hand, and so denies predicates that are true of that case" (O'Neill 1996: 41), whereas abstracting "is a matter of *bracketing*, but not of *denying*, predicates that are true of the matter under discussion" (O'Neill 1996: 40, emphasis in the original). If we base our moral principles on abstract ideas of activity, we can then apply those principles to a broad range of agents regardless of how, beyond those thin abstractions, their capabilities and vulnerabilities differ from one another. Because our principles rely on abstractions rather than idealisations, no agent will be arbitrarily disadvantaged – or advantaged – because of her particular capabilities and vulnerabilities. Conversely, relying on an idealised conception of agents or activity requires a metaphysical justification of these ideals and risks excluding those whose capabilities and vulnerabilities differ from those of the ideal agent.

The method O'Neill proposes for working out whom to accord ethical standing in a given situation without resorting to unvindicated ideals is centred around actions and the presuppositions on which these actions are based. To uncover the relevant presuppositions, O'Neill breaks down our actions into what she argues are the three most fundamental abstractions of human activity: plurality, connectedness, and finitude. Whenever we engage in other-affecting actions, that is, actions that are relevant to justice, we implicitly or explicitly rely on there being others (*plurality*), on there being a connection between us (*connectedness*), and on the fact that agents have limited capacities, capabilities, and vulnerabilities (*finitude*). On her approach, the appropriate scope of one's obligations is determined by the presuppositions about other connected and finite agents on which one's actions rely. She argues that whenever our activity presupposes the existence of finite and connected others, and there is a real possibility our action will bear on them, these others are brought into the scope of ethical consideration (O'Neill 1996: 101). In other words, whenever an agent's activity is based on the presupposition that there are others with limited capacities, capabilities, and vulnerabilities, and there is a real possibility these others be affected by the activity, they must also be included within the agent's scope of ethical consideration.

O'Neill uses this approach to determine the scope of ethical consideration more generally, which she takes to include both justice and virtue. An important difference to what I seek to do in this book is that my investigation will focus only on our duties of justice, more specifically intergenerational climate justice. Earlier in this chapter I explained why it is crucial to show that we have duties of climate justice to future people, more so than other types of moral duties. In what follows, I will therefore not discuss whether we have additional duties of virtue, beneficence, or other, and if so what their scope is, but focus on showing whether, and if so why, we have duties of climate justice to future generations.

On O'Neill's account, the scope of ethical consideration – including justice – includes all those whose existence as agents we acknowledge through our actions and their underlying presuppositions. O'Neill clearly defines these presuppositions as the assumptions of *activity*, not of the agent. Hence, although O'Neill's agents may be unaware of these assumptions or even deny them, they nevertheless commit to them if and when they engage in the relevant activity. A regime's torturing of dissident citizens, for example, presupposes that these are people who can be affected by the government's actions and in turn can affect the regime itself, who are vulnerable to the infliction of pain, and who have the capabilities to impose resistance, even though the tortures may seek to deny their victims' humanity, vulnerabilities, and capabilities (O'Neill 1996: 103; see also 102, note 15). Once they commit to them through their actions, agents cannot coherently deny these assumptions when it comes to fixing the scope of their obligations. That is, by engaging in any activity they are also committing themselves to granting ethical standing to those whom their actions presuppose (O'Neill 1996: 99–102).

O'Neill's is not primarily an account of intergenerational justice, yet it offers a particularly interesting starting point for thinking about questions of justice for future generations. O'Neill herself believes that her arguments should apply in the intergenerational context, too, provided our actions are based on sufficiently strong presuppositions about the future. Given our actions' immense influence on future lives, of which climate change is an especially poignant example, she also acknowledges the urgency of extending our accounts of justice to the intergenerational sphere (O'Neill 1996: 115). One way in which O'Neill believes activity can, and does, acknowledge future people is whenever agents know that their actions' effects will persist in time and will continue to be felt by future generations (O'Neill 1996: 117).

The account of the scope of climate justice I am going to develop in this book will connect to, and in places complement, O'Neill's brief excursion into intergenerational justice. Most importantly, in Chapters 2 and 3 I will provide a detailed account of the ways in which our actions acknowledge future people through the presuppositions of activity and of the implications this has for our duties of justice to them. These intergenerational actions – actions that rely on presuppositions about future people – will form the backbone of my account of scope. Throughout my investigation, I am going to rely on a definition of presuppositions that is very close to O'Neill's. The presuppositions that, I argue, ground duties of justice are presuppositions about others as agents, that is, about others with certain capabilities or vulnerabilities. Like O'Neill's, these are presuppositions of the action in question and not a state of mind of the agent. This means that while they must shape the action, they do not need to be acknowledged by the agent in order to ground moral duties. They are best described as *preconditions* of the action in its current form. In other words, these presuppositions about others explain certain features of the action and render it conceptually coherent. Because these presuppositions are a feature of the agent's action rather than her consciousness, the agent does not need to have acknowledged or even to be aware of them for them to be the kind of presupposition that can ground duties of justice. To underline the fact that the presuppositions which I refer to are not related to the agent's state of mind, throughout this book I will refer to them only as *presuppositions* and not, like O'Neill, as the *assumptions* of activity.

As an example, consider the action *Ann handing Joe a book*. This activity presupposes that Joe is able and at least likely to take hold of the book without dropping it on the floor. That is, it presupposes that the book will reach Joe and that he has the necessary capabilities to grab it. Whether or not Joe really grasps the book does not make a difference to these presuppositions. That Joe, unbeknownst to Ann, will be shaken by a sudden muscle spasm and drop the book on the ground does not alter Ann's action or what the action presupposes at the moment Ann engages in it. It is merely going to make it an unsuccessful action.

The presuppositions on which our actions are based are a central element of the account of scope I am going to develop. In what follows, I will argue that our actions and their presuppositions give us reasons to include future people within the scope of justice. Because this book aims to develop an account of the scope of climate justice, we must not only show that we have such obligations but do so in a way that meets the intergenerational climate challenge. I am therefore going to construct the account in three steps that tackle three of the challenge's most critical underlying problems head on. In Chapters 2, 3, and 4 I am going to address the problems of uncertainty, of moral complexity, and of how to include future people within the scope of our obligations. Those are three crucial issues that any moral response to climate change must grapple with. I will argue that these problems ground three necessary requirements which an account of scope must fulfil in order to meet the intergenerational climate challenge.

Unlike O'Neill, throughout my investigation I will work with a thin yet substantive account of the basic requirements of justice. I will argue in later chapters that these basic needs are most helpfully conceptualised in terms of the conditions necessary to live an autonomous life, or autonomy conditions. The conditions of autonomy are an abstraction of what we need in order to live and act as autonomous agents and which – as I will argue in Chapters 2 and 3 – we therefore and for the sake of coherence owe to those whom our actions presuppose. Relying on a substantive account of the basic requirements of justice allows my account to distinguish between those obligations that give rise to duties of justice – because they affect, are predicted to affect, or risk affecting the basic requirements of justice – and those that may ground other types of moral obligations. It is important to note that the conditions of autonomy are an abstraction, and not an idealisation, of agents and their activity. Agents are not required to possess any autonomy conditions in order to fall within the scope of our obligations. Rather, the conditions of autonomy are a measure of those goods or states which allow agents to act – regardless of the capabilities and vulnerabilities they already possess – and therefore ought, as a matter of justice, to be provided to all connected and finite agents.

I am going to assume that in order to successfully ground obligations of justice in the face of the intergenerational challenge an account of scope must respond to these criteria in a way that is both *action-guiding* and *normatively accurate*. A principle or theory is action-guiding if it provides the agent with useful practical guidance for the action under consideration (Burri, forthcoming). That is, its guidance must be clear enough and accessible to the agent. This is especially important for principles of climate justice, as they are aimed at providing practical guidance in response to an actual and urgent problem. Because we are concerned with acting justly in response to climate change, we should also want principles of climate justice to be normatively accurate. By that I mean that principles of climate justice should correctly identify the morally relevant features of an action.[15] The account of scope I am going to formulate will be built according to the three identified criteria as well as the need to provide an action-guiding and normatively accurate response to them. It will therefore, so I will argue, successfully show that and why we have intergenerational obligations of climate justice despite the intergenerational challenge.

A framework for intergenerational climate justice

We face an unavoidable degree of uncertainty when predicting the future impacts of climate change. Contrary to situations of pure risk that allow us to assign probabilities to possible outcomes, in situations of uncertainty we can predict the possible effects of our actions but not the likelihood of each outcome. That we are uncertain about future climate change is one of the main problems the intergenerational climate challenge calls our attention to. Some uncertainty is unavoidable for any prediction we make about the future, but in the case of climate change it is compounded by the complexity of the climate system and the unpredictability of, for example, the existence of tipping points that would lock in especially dangerous and large-scale climate changes. This means that we may not always know with sufficient certainty what the future impacts of our actions are going to be. Most actions related to climate change, however, nevertheless have foreseeable effects on the climate. That is, when we engage in activity related to climate justice it is in most cases foreseeable to us that our actions are going to negatively impact the climate and in turn future people, even if we cannot assign exact probabilities to these future events.

In Chapter 2, I am going to argue that the uncertainty about the future impacts of our actions is one of the main problems we face when fixing the scope of climate justice and that an account of scope must be able to deal with this in order to meet the intergenerational challenge. I will introduce the first criterion for a successful account of scope, C1:

> *DEALING WITH UNCERTAINTY. An account of scope must be able to accommodate uncertainties about the future effects of our climate changing and mitigating activities.*

I am also going to show that whether or not the potential negative impacts of our actions are foreseeable to us makes an important difference to whether those predictably affected by them fall within the scope of our obligations. I will argue that an account of scope can use the presuppositions on which foreseeably risky actions are based to determine the scope of agents' obligations of climate justice.

In Chapter 3, I am going to address the problems that the sheer complexity of climate change poses for an account of the scope of climate justice. I will introduce the second criterion that an account of the scope of climate justice must meet, C2:

> *DEALING WITH COMPLEXITY. An account of the scope of climate justice must be able to respond to climate change as a complex problem of justice.*

The issue of complexity gets to the heart of the intergenerational challenge, as it highlights how the number of diverse problems which make up the challenge is itself a hurdle for the moral evaluation of climate change. When thinking about climate change from a moral perspective we have to account for the temporal and spatial dispersion of its causes and effects, the severity of its impacts and their dependence on existing levels of vulnerability, and our uncertainty about how and when these impacts will eventuate, and we must do so all at once. The interaction of these issues with one another makes climate change a particularly complex problem of justice and sets it apart from more common moral issues we are used to deal with.

The difficulty, however, lies not in the complexity itself but therein that we seem to lack the moral norms to adequately respond to it. Stephen Gardiner and

Dale Jamieson have been particularly vocal about this concern. In Chapter 3 I will address their respective arguments. In response, I will show that by grounding our duties to future people on our actions and their presuppositions we can uncover the necessary moral norms, successfully determine the scope of climate justice, and argue for the inclusion of future people in it. In this way we can overcome the problem of complexity, at least with regard to scope-setting.

The third and final question we need to address is how future people are to be included in an account of the scope of climate justice. We need to ask this question in order to understand if there is anything especially morally wrong about passing the buck to future people, and if so what it is. In addition, how we answer this question will set the ground rules for the substance of intergenerational climate justice. I will argue that it makes an important difference to an account of scope and to the resulting substantive obligations of justice whether future people are included for their own sake or because of their derivative importance to us. In turn, the ability of an account of climate justice to fully respond to the intergenerational challenge can be affected by how future people are included in its scope because of the impact this can have on the obligations for which we can later argue. In Chapter 4, I will address this question and argue that our account of scope must value future people for their own sake and as ends in themselves. I will introduce the third and final criterion for a successful account of scope, C3:

> *INCLUDING FUTURE PEOPLE FOR THEIR OWN SAKE. An account of the scope of climate justice must include each person for their own sake and as an end in themselves.*

I will show that if we fail to do so we risk wrongly treating future people as mere means to our ends and using our account of scope to justify unjust future worlds. This possibility alone is strongly reminiscent of the danger of moral corruption Gardiner calls our attention to. Moreover, actually engaging in either behaviour would fail to account for the vulnerability of future people to our actions and thus fall short of meeting the intergenerational challenge.

The discussion in Chapter 4 will be centred around Samuel Scheffler's account of our relation to future generations. In it, he suggests that our dependency on future people gives us additional important reasons to attend to their interests (Scheffler 2013). I will argue that, regardless of our dependency on them, it is important that future people be included in an account of scope for their own sake and not because of their importance to us. Scheffler does, however, provide an especially interesting interpretation of the role of future people in present lives. He argues that many present activities presuppose the existence of future people, as they are actions that are only valuable to us when placed within the context of the ongoing existence of humanity. In Chapter 5 I will pick up this suggestion, arguing that Scheffler's conjecture implies that a very wide range of actions presupposes future people and thus, on my proposed account of scope, triggers duties of intergenerational justice.

This is one of the three issues Chapter 5 is going to deal with. The overall aim of the chapter is to draw the analysis together and formulate a final account of the scope of climate justice which meets all three relevant criteria. The account I am going to propose will rely on intergenerational actions, a coherence requirement, and a normative requirement to show that the scope of an agent's duties of justice includes all those present and future others on whom her actions rely. In turn, if the agent is engaged in activities relevant to climate justice, the scope of her duties of climate justice also extends to those on whom her actions rely. In Chapter 5 I will also clarify three unresolved issues about the scope of climate justice. I will draw on Scheffler's account to more closely define the range of actions that give rise to intergenerational obligations; I will clarify that the proposed account can ground obligations to both near and remote future people; finally, I will elaborate the idea, introduced in Chapter 3, that some of our actions may ground a form of intergenerational cooperation, and assess that cooperation's effects on the scope of our obligations. I will conclude by showing how an account of scope based on the presuppositions behind our actions can allow us to circumvent the non-identity problem and the non-existence challenge.

Notes

1 For more on the distinction between these two ways of thinking about climate justice – harm-avoidance and burden-sharing – see Caney 2014a.
2 See Caney 2008, 2014b for some of these objections.
3 Broome's account of climate ethics is laid out extensively in "Counting the Cost of Global Warming" (1992) and "Climate Matters: Ethics in a Warming World" (2012).
4 For more on the non-identity problem, see Parfit 1984, 1986, 2010, 2017; Page 2008; Woodward 1986.
5 In his account of climate ethics Broome draws a distinction between public and private morality. He denies that public entities such as governments may have duties of intergenerational justice, and one important reason for that is the non-identity problem. However, he believes that individuals may owe duties of justice to future generations, as their individual actions do not have a comparably large effect on who is later born (Broome 2012). Here, I address his argument against a public duty of justice to future generations. This is because in this book I am primarily concerned with the obligations of organised collective agents to address climate change.
6 For arguments for and against such pure time preference, see, for example, Broome 1992, 2012; Caney 2014b: 323–324; Cowen and Parfit 1992; Rawls 1999: 259–260; Parfit 1984. See also Chapter 1; it is generally accepted that claims of justice may however be discounted for other factors such as uncertainty (Wolf 2009).
7 Gardiner (2017) later expands the framework to include a fourth, ecological storm. Because the focus of this project is limited to the human dimension of climate change, I refer to Gardiner's earlier interpretation of the moral storm including only the global, intergenerational, and theoretical storms. The non-human or ecological dimension of climate change brings out distinct concerns which are outside the scope of this book.
8 For a more detailed explanation, see Broome 1994: 137–139.
9 For arguments in favour of discounting future climate-related claims as well as strong counterarguments, see Caney 2008, 2014b; Parfit 1984. It is important to note that not all scholars who frame climate duties in terms of beneficence and deny any obligations of intergenerational justice argue for a positive discount rate for time. The most notable example is Broome 1992, 2012.

10 For additional arguments and possible responses, see Arrow et al. 1995; Caney 2008, especially pp. 548–550; Dasgupta et al. 1999; Gardiner 2011: chapter 8; Parfit 1984: 480–486; Weisbach and Sunstein 2008; Weitzman 1998.
11 As, for example, Nordhaus 2008.
12 That we have intergenerational obligations does not imply they should be fulfilled at the cost of intragenerational obligations of justice; to the contrary, trade-offs between intra- and intergenerational obligations may be required if, or when, both cannot be satisfied at the same time.
13 Though this does not need to be the action's only, or main, aim.
14 This definition of actions draws on Peter Caws' definition of acts, which he argues must be intended and consciously performed by the agent (Caws 1995: 327); and Onora O'Neill's understanding of activity relevant to justice and virtue, which she takes to include "individuable acts and responses, feelings and attitudes, support for policies, and participation in practices" (O'Neill 1996: 99).
15 For a discussion of action-guiding and normatively accurate principles in relation to permissible individual risk-imposition, see Burri (forthcoming).

Reference list

Abizadeh, Arash. 2007. 'Cooperation, Pervasive Impact, and Coercion: On the Scope (Not Site) of Distributive Justice'. *Philosophy & Public Affairs* 35 (4): 318–358. https://doi.org/10.1111/j.1088-4963.2007.00116.x.

Arrow, Kenneth, William R. Cline, Karl-Göran Mäler, R. Squitieri, Joseph E. Stiglitz, and Mohan Munasinghe. 1995. 'Intertemporal Equity, Discounting, and Economic Efficiency'. In *Climate Change 1995: Economic and Social Dimensions of Climate Change, Contribution of Working Group III to the Second Assessment Report of the Intergovernmental Panel on Climate Change*. New York: Cambridge University Press.

Barry, Brian. 1978. 'Circumstances of Justice and Future Generations'. In *Obligations to Future Generations*, edited by Richard I. Sikora and Brian Barry. Cambridge: White Horse Press.

Beckerman, Wilfred, and Joanna Pasek. 2001. *Justice, Posterity, and the Environment*. Oxford: Oxford University Press.

Broome, John. 1992. *Counting the Cost of Global Warming*. Cambridge: White Horse Press.

———. 1994. 'Discounting the Future'. *Philosophy and Public Affairs* 23 (2): 128–156. https://doi.org/10.1111/j.1088-4963.1994.tb00008.x.

———. 2012. *Climate Matters: Ethics in a Warming World*. New York: W. W. Norton & Company.

Burri, Susanne. Forthcoming. 'Conceptualising Permissible Risk Imposition Without Probabilities'. Draft Paper.

Caney, Simon. 2008. 'Human Rights, Climate Change, and Discounting'. *Environmental Politics* 17 (4): 536–555. https://doi.org/10.1080/09644010802193401.

———. 2009a. 'Climate Change, Human Rights and Moral Thresholds'. In *Human Rights and Climate Change, by Mary Robinson*, edited by Stephen Humphreys, 69–90. Cambridge: Cambridge University Press.

———. 2009b. 'Climate Change and the Future: Discounting for Time, Wealth, and Risk'. *Journal of Social Philosophy* 40 (2): 163–186. https://doi.org/10.1111/j.1467-9833.2009.01445.x.

———. 2014a. 'Two Kinds of Climate Justice: Avoiding Harm and Sharing Burdens: Two Kinds of Climate Justice'. *Journal of Political Philosophy* 22 (2): 125–149. https://doi.org/10.1111/jopp.12030.

———. 2014b. 'Climate Change, Intergenerational Equity and the Social Discount Rate'. *Politics, Philosophy & Economics* 13 (4): 320–342. https://doi.org/10.1177/14705 94X14542566.

Caws, Peter. 1995. 'Minimal Consequentialism'. *Philosophy* 70 (273): 313–339. https://doi.org/10.1017/S0031819100065542.

Climate Analytics and New Climate Institute. 2021. 'The CAT Thermometer'. *Climate Action Tracker*. November. https://climateactiontracker.org/global/cat-thermometer/.

Cohen, Andrew I. 2009. 'Compensation for Historic Injustices: Completing the Boxill and Sher Argument'. *Philosophy & Public Affairs* 37 (1): 81–102. https://doi.org/10.1111/j.1088-4963.2008.01146.x.

Cowen, Tyler, and Derek Parfit. 1992. 'Against the Social Discount Rate'. In *Justice Between Age Groups and Generations*, edited by Peter Laslett and James S. Fishkin, 144–161. Philosophy, Politics, and Society. New Haven: Yale University Press.

Dasgupta, Partha, Karl-Göran Mäler, and Scott Barrett. 1999. 'Intergenerational Equity, Social Discount Rates and Global Warming'. In *Discounting and Intergenerational Equity*, edited by Paul Portney and John Weyant. Washington, DC: Routledge.

De George, Richard T. 1979. 'The Environment, Rights, and Future Generations'. In *Ethics and Problems of the 21st Century*, edited by Kenneth E. Goodpaster and Kenneth M. Sayre. Notre Dame: University of Notre Dame Press.

Dennig, Francis. 2018. 'Climate Change and the Re-Evaluation of Cost-Benefit Analysis'. *Climatic Change* 151 (1): 43–54. https://doi.org/10.1007/s10584-017-2047-4.

Elliot, Robert. 1989. 'The Rights of Future People'. *Journal of Applied Philosophy* 6 (2): 159–170. https://doi.org/10.1111/j.1468-5930.1989.tb00388.x.

Gardiner, Stephen M. 2003. 'The Pure Intergenerational Problem'. *The Monist* 86 (3): 481–500. https://doi.org/10.5840/monist200386328.

———. 2004. 'Ethics and Global Climate Change'. *Ethics* 114 (3): 555–600. https://doi.org/10.1086/382247.

———. 2006. 'A Perfect Moral Storm: Climate Change, Intergenerational Ethics and the Problem of Moral Corruption'. *Environmental Values* 15 (3): 397–413. https://doi.org/10.3197/096327106778226293.

———. 2011. *A Perfect Moral Storm*. Oxford: Oxford University Press. https://doi.org/10.1093/acprof:oso/9780195379440.001.0001.

———. 2017. 'The Threat of Intergenerational Extortion: On the Temptation to Become the Climate Mafia, Masquerading as an Intergenerational Robin Hood'. *Canadian Journal of Philosophy* 47 (2–3): 368–394. https://doi.org/10.1080/00455091.2017.1302249.

Herstein, Ori J. 2009. 'The Identity and (Legal) Rights of Future Generations'. *The George Washington Law Review* 77: 1173–1215.

Hume, David. 1987. *Essays, Moral, Political, and Literary [1777]*. Edited by Eugene F. Miller. Indianapolis: Liberty Classics.

IPCC. 2014. *Climate Change 2014: Mitigation of Climate Change: Contribution of Working Group III to the Fifth Assessment Report of the Intergovernmental Panel on Climate Change*. New York: IPCC.

———. 2018. *IPCC Special Report Global Warming of 1.5°C*. Incheon, Republic of Korea: IPCC.

———. 2021. *Climate Change 2021: The Physical Science Basis. Contribution of Working Group I to the Sixth Assessment Report of the Intergivernmental Panel on Climate Change*. Cambridge: IPCC.

Kumar, Rahul. 2003. 'Who Can Be Wronged?' *Philosophy and Public Affairs* 31 (2): 99–118. https://doi.org/10.1111/j.1088-4963.2003.00099.x.

Lomborg, Bjørn. 2001. *The Skeptical Environmentalist: Measuring the Real State of the World*. Cambridge and New York: Cambridge University Press.

Macklin, Ruth. 1981. 'Can Future Generations Correctly Be Said to Have Rights?' In *Responsibilities to Future Generations: Environmental Ethics*, edited by Ernest Partridge, 151–156. Buffalo, NY: Prometheus Books.

Meyer, Lukas. 2003. 'Past and Future: The Case for a Threshold Notion of Harm'. In *Rights, Culture and the Law*, edited by Lukas Meyer, Stanley L. Paulson, and Thomas W. Pogge. Oxford: Oxford University Press. www.oxfordscholarship.com/view/10.1093/acprof:oso/9780199248254.001.0001/acprof-9780199248254.

———. 2016. 'Intergenerational Justice'. In *The Stanford Encyclopedia of Philosophy*, edited by Edward N. Zalta. Stanford: Metaphysics Research Lab, Stanford University. Summer.

Miller, David. 2017. 'Justice'. In *The Stanford Encyclopedia of Philosophy*, edited by Edward N. Zalta. Stanford: Metaphysics Research Lab, Stanford University. Fall.

Nordhaus, William. 2008. *A Question of Balance: Weighing the Options on Global Warming Policies*. New Haven: Yale University Press.

Nussbaum, Martha. 2003. 'Capabilities as Fundamental Entitlements: Sen and Social Justice'. *Feminist Economics* 9 (2–3): 33–59. https://doi.org/10.1080/1354570022000077926.

O'Neill, Onora. 1996. *Towards Justice and Virtue: A Constructive Account of Practical Reasoning*. Cambridge: Cambridge University Press.

———. 2000. *Bounds of Justice*. Cambridge: Cambridge University Press.

———. 2001. 'Agents of Justice'. *Metaphilosophy* 32 (1–2): 180–195. https://doi.org/10.1111/1467-9973.00181.

Page, Edward A. 2008. 'Three Problems of Intergenerational Justice'. *Intergenerational Justice Review* Special Edition: *Groundworks for Intergenerational Justice* 2: 9–12.

Parfit, Derek. 1984. *Reasons and Persons*. Oxford: Oxford University Press.

———. 1986. 'Comments'. *Ethics* 96 (4): 832–872.

———. 2010. 'Energy Policy and the Further Future: The Identity Problem'. In *Climate Ethics: Essential Readings*, edited by Stephen M. Gardiner. Oxford: Oxford University Press.

———. 2017. 'Future People, the Non-Identity Problem, and Person-Affecting Principles'. *Philosophy & Public Affairs* 45 (2): 118–157. https://doi.org/10.1111/papa.12088.

Posner, Richard A. 2004. *Catastrophe: Risk and Response*. Oxford: Oxford University Press.

Rawls, John. 1999. *A Theory of Justice*. Revised Edition. Cambridge, MA: Belknap Press of Harvard University Press.

Rivera-López, Eduardo. 2009. 'Individual Procreative Responsibility and the Non-Identity Problem'. *Pacific Philosophical Quarterly* 90 (3): 336–363. https://doi.org/10.1111/j.1468-0114.2009.01344.x.

Scheffler, Samuel. 2013. *Death and the Afterlife*. Edited by Niko Kolodny. Oxford: Oxford University Press.

Sen, Amartya. 2004. 'Elements of a Theory of Human Rights'. *Philosophy & Public Affairs* 32 (4): 315–356. https://doi.org/10.1111/j.1088-4963.2004.00017.x.

Stern, Nicholas, ed. 2007. *The Economics of Climate Change: The Stern Review*. Cambridge: Cambridge University Press.

United Nations Climate Change. 2015. 'Paris Agreement to the United Nations Framework Convention on Climate Change'. https://unfccc.int/sites/default/files/english_paris_agreement.pdf.

———. 2021. 'Glasgow Climate Pact'. https://unfccc.int/sites/default/files/resource/cop26_auv_2f_cover_decision.pdf.

United Nations Environment Programme (UNEP). 2015. *The Emissions Gap Report 2015*. Nairobi: UNEP.

Valentini, Laura. 2013. 'Justice, Charity, and Disaster Relief: What, If Anything, Is Owed to Haiti, Japan, and New Zealand?' *American Journal of Political Science* 57 (2): 491–503. https://doi.org/10.1111/j.1540-5907.2012.00622.x.

Velleman, J. David. 2008. 'The Identity Problem'. *Philosophy & Public Affairs* 36 (3): 221–244. https://doi.org/10.1111/j.1088-4963.2008.00139_1.x.

Weisbach, David A., and Cass R. Sunstein. 2008. 'Climate Change and Discounting the Future: A Guide for the Perplexed'. *SSRN Electronic Journal*. https://doi.org/10.2139/ssrn.1223448.

Weitzman, Martin L. 1998. 'Why the Far-Distant Future Should Be Discounted at Its Lowest Possible Rate'. *Journal of Environmental Economics and Management* 36 (3): 201–208. https://doi.org/10.1006/jeem.1998.1052.

Wenar, Leif. 2017. 'John Rawls'. In *The Stanford Encyclopedia of Philosophy*, edited by Edward N. Zalta. Stanford: Metaphysics Research Lab, Stanford University. Spring.

Wolf, Clark. 2009. 'Intergenerational Justice, Human Needs, and Climate Policy'. In *Intergenerational Justice*, edited by Axel Gosseries and Lukas H. Meyer. Oxford: Oxford University Press.

Woodward, James. 1986. 'The Non-Identity Problem'. *Ethics* 96 (4): 804–831. https://doi.org/10.1086/292801.

2 Dealing with uncertainty

Justice and future climate changes

Being able to cope with uncertainty is especially important for an account of climate justice, since – although it is virtually certain that climate change will have a range of severely negative impacts on generations to come – even the most accurate climate models leave us with large uncertainties about what changes will happen when, and exactly how bad their impacts will be. This means we have to devise principles of climate justice and make policy decisions under conditions of risk, uncertainty, and ignorance. What matters for the purpose of this book is that uncertainty about future climate changes also challenges our ability to set the scope of climate justice.

In order to successfully respond to the intergenerational climate challenge, an account of the scope of climate justice must therefore meet the following criterion C1:

> DEALING WITH UNCERTAINTY. An account of scope must be able to accommodate uncertainties about the future effects of our climate changing and mitigating activities.

Two examples can help illustrate how uncertainty may affect the scope of our obligations. Imagine, first, the following:

> ASTEROID DOOMSDAY RISK. This is a risk of unknown magnitude that, at some point in time, an asteroid will strike the earth, causing significant destruction of the natural environment and infrastructure, as well as a significant number of deaths. We have reasons to believe this risk exists but are unable to specify it much further. We know that asteroids orbit the sun and that if their orbits were to intersect with that of the earth it would come to a collision. However, we do not have any evidence suggesting that any of our actions are affecting the magnitude of the risk or its causal mechanisms. What we do know is that we can take precautions against the risk of a collision by investing large sums of money in impact avoidance strategies.

Asteroid doomsday is clearly a significant risk, but it is also a very diffuse one: we are uncertain about its magnitude and, at the same time, lack evidence to show

DOI: 10.4324/9781003258902-2

whether we are acting in ways that are increasing the risk of impact. This means we are unable, for example, to get a sense of its magnitude by evaluating the extent to which we are or are not engaging in actions that increase this specific risk. Also, we have limited grounds on which to judge whether our precautions will in fact serve to minimise the risk of impact. I have the intuition that, primarily because of the level and type of uncertainty just outlined, it would not be reasonable to hold a set of people, a certain generation, or a set of generations at fault for imposing the risk of impact on others. In other words, it seems plausible to think that we cannot attribute a duty of justice to any one agent or agents requiring them to invest scarce resources in the prevention of the *asteroid doomsday risk*.

This should not be misunderstood for an excuse to pass the buck of asteroid protection to future generations: it seems equally plausible that we do have a less stringent duty of beneficence to use some of our resources to protect us and future generations against the risk of collision. What I am suggesting is that if we do have moral reasons to invest in asteroid protection, these are not tied to considerations of justice, so if competing for the same resources, under current circumstances priority should be given to other justice-based obligations.

As a second example, consider the following (actually occurring) risk:

> *MELTING OF THE WEST ANTARCTIC ICE SHEET (WAIS). There is a risk the WAIS will melt completely, which would increase sea levels by approximately an additional 10 feet and have catastrophic impacts on coastal populations (Hartzell-Nichols 2017). Although we are already seeing a significant degree of melting, there is still a lot of uncertainty about the rate and timing at which melting will continue in the future and about its impacts on future generations (Guy 2019; Hartzell-Nichols 2017; Shankman 2019). Just as in the asteroid doomsday scenario, we cannot assign an exact probability to the risk of melting. Yet unlike the asteroid case, we know that our actions are increasing the risk of melting: we know the ice sheet is melting due to rising sea and air temperatures which are caused by an accumulation of greenhouse gases in the atmosphere, which in turn is emitted as a by-product of many human activities. We also know that by limiting our emissions of greenhouse gases we can stop or at least limit the extent to which we are contributing to increasing this risk. Moreover, we are able to control our emissions and could take the necessary precautions at no excessive cost to ourselves.*

Intuitively, it seems plausible to think that in this case we do have an obligation of justice to prevent the risk from eventuating, despite the uncertainty involved in defining the risk. I believe this difference in judgement may be due to the level and type of uncertainty involved in each case. Unlike the *asteroid doomsday risk*, in the *WAIS* case we know that our actions are interfering in this risk in ways that increase it. On the one hand, this means that – because we know how much greenhouse gas has been and is still being emitted – we have more information we can use to get a sense of the magnitude of the risk and how urgently we need to address it. On the other hand, it also seems that because we have this additional knowledge, we as a generation could be held at least partly at fault if the risk were to eventuate.

If my intuitions are correct and our degree and type of uncertainty are relevant factors in determining whether a risk and those exposed to it fall within the scope of justice, we need to determine exactly how they do so. We need to know which risks give rise to obligations of climate justice, whose interests must be considered when devising principles to deal with those risks, and why. The purpose of this chapter is to propose a way in which an account of scope can answer these questions in an action-guiding and normatively accurate manner.

In what follows, I will argue that an account of scope can successfully respond to the uncertainty of climate change predictions if (i) it takes each risk-imposition and its justification as the subject of moral evaluation and (ii) it takes risks to give rise to obligations of climate justice if they meet the following conditions: they are foreseeable by a reasonable agent at the time of acting, and they expose others to risks to their autonomy-relevant conditions.

I will begin in the following section by surveying the kinds of risk we face in the context of climate change. Then, I will outline how a common account of scope – the universal consequentialist account – could respond to these challenges. I will argue that this approach is overinclusive because it cannot differentiate between intuitively different categories of risks, and thus fails to be action-guiding in situations of multiple uncertainties; I will also argue that it is underinclusive, and thus normatively inaccurate, as it fails to take account of morally relevant aspects of our actions. I will then propose an alternative approach to scope, which I will argue can produce both action-guiding and normatively accurate results and is therefore better suited to meet the climate challenge. I will also introduce and later flesh out the concept of foreseeability I am going to rely on. Before concluding this chapter, I will revisit the consequentialist account and apply the notion of foreseeability to it; I will argue that even in this amended version, it fails to produce normatively accurate results.

Risk, uncertainty, and ignorance in the context of climate change

Theoretical overview

Our ability to predict future climate changes is marred by risks, uncertainties, and ignorance. In this chapter, I will be mostly concerned with situations of risk and uncertainty. A situation of *risk* is one in which we can measure and assign probabilities to possible outcomes. We act under conditions of risk, for example, when we purchase a lottery ticket for which we know the winning odds.

Uncertainty, on the other hand, describes situations in which we are aware of possible outcomes – for example, because we understand the mechanisms by which an outcome occurs, and know that the conditions for its realisation are accumulating – but in which we are unable to assign probabilities to them (Shue 2010, 2015; see also Ekeli 2004; Hartzell-Nichols 2017; Knight 2002). This applies, for example, to playing the lottery when the odds are unknown.

Situations marked by *ignorance*, conversely, involve so-called unknown unknowns. These are instances in which we cannot predict the possible outcomes of an action, let alone assign probabilities to them (Ekeli 2004). If, for example, customers were to be automatically and unknowingly entered into a lottery whenever they make a purchase at the grocery store, they would be acting under conditions of ignorance.

Because we cannot, by definition, know what we are ignorant about, the question of how to behave justly in situations of ignorance raises a set of difficult but distinct questions for morality, which I will set aside for the purpose of this book. Nonetheless, it is important to note that situations of ignorance are a source of concern for climate justice and policy. While I do not discuss the problems posed by ignorance in greater detail, some of my later arguments will relate to issues of ignorance, too. In particular, I am going to suggest that an agent may have a duty to obtain as much information about the consequences of her actions as can reasonably be expected of her, which will in turn bear on her degree of ignorance. I will return to this point later in the chapter.

Empirical overview

One thing we can say for certain about climate change is that, to a greater or lesser degree, "anthropogenic forces will causally contribute to harmful outcomes via climate change" (Hartzell-Nichols 2017: 8). All our decisions about how to respond to the threat of climate change thus involve a trade-off between the interests in the continued emission of greenhouse gases through climate changing activities and the interests in not being exposed to the harmful effects of climate change.

These opposing interests are held, for the most part, by different sets of people. Consider, for example, a business as usual (BAU) scenario. BAU means that we keep emitting greenhouse gases at the current rate. The fossil fuel economy seems, at least on the surface, to promise greater economic growth in the immediate future than shifting to the greener and less consumerist economy we need in order to prevent dangerous climate change. In addition, ceasing or even altering the emitting activities that are so deeply embedded in our economies and societies would be initially costly for present generations. It is, at least in these respects, in the interest of a large share of present people to continue on a BAU trajectory. On the other hand, it will take centuries if not millennia to reverse the anthropogenic changes caused by greenhouse gases emitted now (IPCC 2013: 28). Because of the temporal element of climate change, the costs of present emissions – in terms of exposure to harmful changes in the climate – will fall largely on future people.[1]

However, despite this relative certainty, we know far less about the precise magnitude and timing of the specific outcomes of our actions. The climate system and the mechanisms that link human activities to climate changes are extraordinarily complex, and risk and uncertainty permeate much of our knowledge about the future effects of climate change (Hartzell-Nichols 2017; see also IPCC 2014a).[2] What complicates matters even further is that because we are dealing with a natural environment, the probabilities of possible outcomes can rarely be pinpointed as accurately as in a lottery. This can blur the lines between situations of risk and situations of uncertainty, so that often, even what we call risks are in fact better characterised as uncertainties.

The Intergovernmental Panel on Climate Change (IPCC), for example, defines a set of "emergent risks" and attributes a probability range to many of its predicted outcomes. At the same time, it also uses measures of confidence – ranging from very low over low, medium, and high to very high – to express the remaining level of uncertainty about the probabilities assigned to these predictions (Oppenheimer et al. 2014; IPCC 2014a). For instance, according to its fifth Assessment Report of 2014, the global mean surface temperature change for 2016–2035 relative to 1986–2005 will *likely* (66–100% probability) be between 0.3°C and 0.7°C, which is expressed with *medium confidence* (IPCC 2014c: 58).

Because the language used is that of risks and possible outcomes are assigned a probability, the case may seem to resemble our earlier example of a lottery with known odds and thus a situation of risk. At the same time, the assigned probabilities can only be expressed as a range, *and* an additional level of uncertainty is added by the measures of confidence. Because of these complicating factors, the distinction between risk and uncertainty is less clear-cut. Ultimately, however, according to our definition the situation is for these reasons better characterised as one of uncertainty rather than one of pure risk (see also Hartzell-Nichols 2017: 8–12).[3]

Modelling future climate changes and their impacts also involves a range of much greater uncertainties which cannot be assigned any useful probability ranges at all. One such example is the melting of the WAIS. While recent findings suggest it may now be melting inevitably, our ability to predict the timing and speed at which this will happen is curtailed by a significant degree of uncertainty (Bamber et al. 2019; Hartzell-Nichols 2017: 86; Shue 2010). These uncertainties often involve the possibility of particularly harmful outcomes, such as the collapse of the WAIS and the associated rise in sea level. It is therefore particularly important that an account of climate justice be able to account for these, too.

Uncertainties, rather than risks, thus dominate our predictions of future climate changes. They also present the more serious challenges for both moral and political assessment, as we are unable to factor probability calculations into our decisions on how to act. Unless otherwise specified, in what follows I will therefore focus on situations of uncertainty, or uncertain risks, understood as situations in which possible outcomes can be assigned only a range of probability or none at all. Where uncertainty is to be distinguished from risk, I will speak of risks as pure risks.

The consequentialist approach to scope

The uncertainties that permeate our understanding of climate change can cause difficulties when it comes to fixing the scope of climate justice. These challenges are particularly visible in consequentialist approaches to climate justice and most so in the universal maximising consequentialist view of scope. This approach to scope is at the basis of, for example, classic utilitarian moral theories. In this section, I will describe this view of scope and assess its ability to cope with situations of uncertainty such as climate change. I will argue that although this particular consequentialist account is widely used as the basis for climate policies in the form of cost-benefit analyses, it is neither action-guiding nor normatively accurate enough to overcome the intergenerational climate challenge. In the following section, I am going to suggest what I take to be a more successful way to account for uncertainty when fixing the scope of climate justice.

Consequentialism as a general approach to morality encompasses a heterogeneous group of moral theories. These theories vary by how they value the consequences of actions or obedience to rules, in other words by their theory of the good and by how they evaluate risks and foreseeability. In this book, I am going to look at one particular consequentialist theory – classic utilitarianism – as a paradigm example of a universal, actual, and maximising form of consequentialism. More specifically, I am interested in the way in which the scope of utilitarian principles is fixed and whether this consequentialist account can satisfactorily answer my question about the scope of climate justice. To recall, I am interested in clarifying whether we have to extend considerations of justice to future people in our actions related to climate justice and if so on what grounds. Importantly, the principles we adopt to answer this question should yield action-guiding and normatively accurate results in situations of uncertainty.

Classic utilitarianism as a moral theory rests on a combination of claims, including claims about scope and its theory of the good. It is a form of evaluative act-consequentialism, according to which moral rightness depends on the value of the consequences of our actions and follows a hedonistic theory of the good, according to which the value of consequences depends only on the pleasure and pain caused (Hooker 2002; Sinnott-Armstrong 2019). We can set some of these claims aside for the moment. The claims that are relevant to my arguments relate to the theory's scope and the range of consequences it considers. Classic utilitarianism is a universal consequentialist account, according to which "moral rightness depends on the consequences for *all* people or sentient beings (as opposed to only the individual agent, members of the individual's society, present people, or any other limited group)" (Sinnott-Armstrong 2019).[4] It also holds that the moral value of an act depends only on its actual consequences and not, for instance, on its foreseen or intended consequences (Hooker 2002; Sinnott-Armstrong 2019). In policymaking, this type of consequentialist thinking is often at play in cost-benefit analyses. Cost-benefit calculations still play a large role in climate-political decision-making so that climate policies can often be influenced by the underlying consequentialist reasoning (see Gardiner 2011: chapter 8; see also Hartzell-Nichols 2017: chapter 4).

Given its underlying universal understanding of scope, a utilitarian consequentialist would answer my question about the scope of climate justice as follows: yes, we do have to consider future people in our actions related to climate justice, and the reason for this is that the rightness of our actions depends on the consequences of our actions, including their consequences for future people. In what follows I will argue that, despite its seemingly clear response, this consequentialist approach fails to yield action-guiding and normatively accurate results in situations of uncertainty. It should therefore be rejected as a method for fixing the scope of climate justice, and we must look for an alternative path to meet the climate challenge.

Other forms of consequentialism – for example, expected consequentialism or rule consequentialism – can rely on different accounts of risk and foreseeability and differ from classic utilitarianism with regard to the consequences they evaluate (see Hooker 2002). Because consequentialist theories are so diverse, what I will argue in this chapter will not, and is not intended to, apply to all of them. Nor do my arguments imply that all consequentialist theories are equally unable to deal with uncertainty in an action-guiding and normatively accurate manner. My arguments are specific to actual, universal consequentialist theories – such as classical utilitarianism – and, more specifically, their approach to scope. Throughout the remainder of this book, when I speak of consequentialism I will therefore be referring to this particular form of consequentialism.

Overinclusiveness

The consequentialist approach can be criticised on two distinct but interrelated grounds. It is on the one hand *overinclusive*; that is, it requires us to consider possible duties of justice where in fact we have none. Because of its overinclusiveness, it cannot adequately distinguish between cases that do and cases that cannot give rise to duties of justice and therefore fails to be action-guiding whenever we face more than one uncertainty. This failure is partly due to the fact that the consequentialist approach is, at the same time, *underinclusive*. Because it focuses solely on consequences, it does not pick up other morally relevant differences about the character of our actions. This, in turn, also limits the extent to which the account can be normatively accurate.

Let me unpack these objections in turn, starting with its overinclusiveness. Because the scope of consequentialist obligations extends to all people, the consequentialist account requires us to consider the consequences of all our actions – including all risk-imposing actions – as potential sources of moral obligations to others. This is so regardless of the extent to which we know whether or how our actions are affecting these risks. Yet in predicting the consequences of our actions we suffer from significant epistemic limitations. This is a broader problem that actual consequentialism faces with regard to its substance and not merely its scope and that Lenman has famously termed "the problem of cluelessness." It lies therein that we are unable to foresee, and therefore take into account, all the causal ramifications, long-term consequences, or side effects of our actions (Lenman 2000). Because consequentialism tells us that the rightness of an action is determined by the goodness of its consequences, the inability to foresee the consequences of our actions limits our ability to morally evaluate them.

One of the most promising consequentialist replies to Lenman's arguments comes from Hilary Greaves. She argues that the cluelessness worry can be overcome by adjusting the focus of consequentialist theories to judge not the objective betterness of acts but their subjective betterness. This makes an important difference, for when judging the subjective betterness of acts, agents are only permitted to assess the foreseeable consequences of their actions. In turn, in cases of simple cluelessness agents will know what the relevant consequences are and be able to evaluate these (Greaves 2016). While a detailed discussion of her proposition and of what the cluelessness problem means for consequentialist theories is beyond the scope of this book, Greaves' solution seems plausible.

What matters for our discussion is that, nonetheless, Greaves concedes that cluelessness remains a problem for consequentialist theories in cases of complex cluelessness. Whereas in cases of simple cluelessness agents are justified to have equal credence in, on the one hand, the unforeseen consequences of their actions being overall good and, on the other hand, their unforeseen consequences being overall bad, cases of complex cluelessness are situations in which there are complex and reasonable arguments suggesting that the unforeseen consequences of an action could be good, and equally complex and reasonable arguments suggesting the opposite. A government deciding whether to go to war and an individual choosing which job offer to accept would, for example, fall under the complex category (Greaves 2016: 334). In these cases, Greaves argues that agents are not justified to have equal credence in the unforeseen consequences of their actions being overall good or overall bad. They are therefore not justified in limiting their assessment to the subjective betterness of acts, or to their foreseen consequences, and in turn remain clueless about how to morally evaluate these actions.[5]

On any plausible account of what a complex and reasonable argument is, actions related to climate change clearly represent cases of complex cluelessness. On the one hand, the uncertainty of climate predictions means we face risks that we know of but to which we cannot assign sufficiently accurate probabilities. On the other hand, our uncertainty extends to the existence and functioning of feedback mechanisms within the climate system, which if triggered may significantly amplify the causal ramifications of our actions. All of this greatly limits our ability to predict and evaluate the consequences of our actions in the context of climate change.

Yet despite our cluelessness, if, as the consequentialist account stipulates, the scope of our theory is to extend to all beings, we must equally consider all our risky actions as potential sources of duties to future people.

Recall our asteroid and WAIS examples. In both cases we face an uncertain future, yet in the *WAIS* case we know that our actions are interfering in the risk. We understand that our actions are increasing the risk, even if not by how much exactly. We are also able to limit or end our interference in the risk at not too high a cost to ourselves. In the *asteroid doomsday* case, we understand that and why there is a risk of a collision yet have no evidence indicating that our actions are in any way influencing this risk. I suggested that it would be unreasonable to believe any set of people to be at fault for a risk as diffuse as the *asteroid doomsday risk* and that this suggests that we cannot have an obligation of justice to future people to prevent it. Conversely, the opposite seems to be true for the *WAIS risk*.

The universal consequentialist view of scope fails to make this distinction, as the risk-exposed in both cases fall equally within the scope of our principles. This means that both cases are to an equal extent a possible source of obligations, and both their consequences must be carefully weighed to judge whether our actions do in fact give rise to such obligations. This is so despite an important difference in the extent to which we can gauge the effect our actions have on each risk. If my intuition is correct and *asteroid doomsday* and *WAIS* warrant different responses from a theory of justice, this difference should be reflected when fixing the scope of our obligations.

The first problem the consequentialist account faces is that, insofar as it fails to distinguish between these types of uncertainty, it can be criticised for being *overinclusive*. The account is overinclusive as it requires agents to consider all risk-exposed others as falling within the scope of their obligations, irrespective of other possibly relevant and differentiating factors about each risk-imposing act. This raises a problem for our task of formulating an action-guiding account of scope as it limits the extent to which the account can offer useful practical guidance to agents. Because it requires agents to consider all risk-exposed others as falling within the scope of their obligations without further differentiating between actions, it will be less apt at providing practical guidance than an account of scope that draws a more refined distinction between cases such as *asteroid doomsday* and *WAIS*. Because on the consequentialist account all risky actions can equally be the source of duties of justice, including cases like *asteroid doomsday*, in which we lack an understanding of whether our actions are interfering in the risk and are thus unable to even approximate its likelihood, this approach will likely fail to be action-guiding in situations with multiple uncertainties. In other words, it will fail to give clear enough and accessible practical guidance to an agent who wants to act justly in situations where multiple uncertainties compete for scarce resources.

The upshot of this objection is not that we can never have obligations of justice to people exposed to uncertain risks. It is rather that, in addition to the possible consequences of what we do, there are likely to be other morally relevant considerations about actions under uncertainty that affect whether acts give rise to duties of justice or not.

Underinclusiveness

In addition to its problem of overinclusiveness, the consequentialist account also fails to provide accurate guidance on the scope of our obligations in a case like *WAIS*, about which we know more than *asteroid doomsday*, but which still counts as a case of complex cluelessness. This is because the consequentialist account faces a second problem of being *underinclusive*. Its underinclusiveness is at the heart of its cluelessness problem and partly explains why the account is overinclusive.

The charge of underinclusiveness relates not to a failure to include instances of risk-impositions or risk-exposed others but to the lack of consideration for other morally relevant aspects of our actions. In particular, the consequentialist account of scope can be criticised for failing to consider the character of our actions when fixing the scope of our duties in situations of uncertainty. The character of our actions, however, seems to be relevant to the moral evaluation of actions under uncertainty, as it can explain our different intuitions in the *asteroid doomsday* and *WAIS* scenarios.

In the remainder of this section, I will argue that the character of our actions, particularly the presuppositions on which our actions are based, explains why we have duties of justice in certain cases of uncertainty but not others, thus allowing us to determine the range of others that fall within the scope of our obligations.

The character of our actions includes not only the agent's behaviour in performing the action, that is, what is visible to others, but extends to the presuppositions on which her actions are based. Other-affecting activity is necessarily based on presuppositions about how others may be affected by or react to it, in other words about the capabilities and vulnerabilities of these others. Bill hiding his candy stash under his pillow, for example, relies on the presupposition that there are other agents – his sister and parents – who have a sweet tooth and may eat his candy if left in plain sight. Recall that, because these are presuppositions of activity rather than the agent's consciousness, Bill can but does not need to acknowledge or be aware of his action's presuppositions for them to be relevant. Yet these presuppositions are critical in shaping our actions. The underlying presuppositions about others explain why Bill hides his candy rather than storing it more appropriately in a cupboard.

Giving more consideration to the presuppositions of our actions rather than just their consequences can do two things to help with the moral assessment of risks and uncertainties. On the one hand, shifting the focus of moral evaluation onto the character of our actions means we have better epistemic access to the information we need in order to evaluate them. We are therefore considerably less clueless about the moral status of our actions. On the other hand, zooming in on the difference in the character of our actions can explain the difference in our intuitive responses to the *asteroid doomsday* and *WAIS* risks. To the extent that it does explain this difference, the character of our actions thus seems to be morally relevant to the scope of our obligations.

The character of risk-imposing actions can make a morally relevant difference as follows. It is the character of our actions that exposes the *WAIS* risk, but not the *asteroid doomsday* risk, as a particularly problematic kind of risk-taking from the point of view of justice. The risk we face in the *WAIS* case has the following features: we understand that our actions are affecting the risk in a way that increases it, and we have the knowledge and ability to disrupt this interference.[6] We can call risks that meet these conditions *foreseeable* risks. I will return to this term in the following sections to provide a more detailed definition of what I take it to mean in this context, as well as an explanation of why I take it to be an important concept.

Although we do not know for certain which future scenario will eventually unfold, the suggested conditions of foreseeability give us reasonable grounds to hold that failing to take precautions against the melting of the ice sheet, that is, failing to limit our greenhouse gas emissions, will expose future people to the risk of substantial harm. Knowing that our actions are affecting the risk as well as the extent to which we are engaging in these actions allows us to get a sense of the magnitude of the risk and to know that the risk of substantial harm to future people is increasing. This includes the risk of harm to fundamental human necessities, such as access to the means of physical subsistence and the ability to move freely.

Risk-imposing present people, on the other hand, could limit their emissions at no excessive cost to themselves and without threatening their own ability to meet fundamental human needs (Green 2015; Stern 2007). Yet decreasing their emissions would, at least initially, impose non-negligible financial costs on present generations. Hence, while a BAU scenario with little to no emission cuts imposes substantial risks on future people, this option seems to be in the interest of those present people who engage in emitting activity for it does not expose them to risks or costs.[7]

Against this background, the risk-imposing actions of present people take on a specific character. Given the foreseeability of the risk they are contributing to, our emitting activities seem to be based on certain presuppositions about future people and the weight of their interests relative to ours. When actions impose foreseeable risks, the underlying presuppositions of activity will include a weighing of the interests of the risk-imposing agents against those of the risk-exposed others. The continued pursuit of a risk-imposing BAU trajectory by present generations suggests that, in this case, more weight is given to the interests of present people in avoiding mitigation costs over the interests of future generations in maintaining their ability to meet their fundamental needs.

In other words, imposing a foreseeable risk of significant harm on others where this could be avoided at a not-excessive cost to oneself relies on the presuppositions that, on the one hand, the future risk-exposed will be vulnerable to the risk in certain ways and, on the other hand, that the interests of the risk-exposed are to be weighed less than one's own. This is relevant to considerations of intergenerational justice for it shows that the action in question is based on presuppositions about future others possessing certain vulnerabilities and capabilities. I will argue in the following section that these presuppositions about others are a useful starting point for fixing the scope of our obligations under uncertainty, as for the sake of coherence we ought to consider these others in our principles of justice, too. For the moment, it is important to note only that this fact about the character of our actions – whether or not they are based on presuppositions about the capabilities of others – seems to be of moral relevance to the evaluation of our actions under uncertainty.

This fact about the character of our actions can also help us differentiate between our two examples where the consequentialist account failed to do so. In the *WAIS* case, our actions are necessarily based on presuppositions about future people as we are knowingly acting in a way that increases the risk of substantial harm to them. In the *asteroid doomsday* case, the character of our actions differs. Although we know that the risk of a collision exists, in addition to being uncertain about its magnitude we have no evidence suggesting that our actions are interfering in the risk. That means that we cannot, by looking at our actions, gauge the order of magnitude of the risk or whether it is likely to be increasing or decreasing. It also means that we are not knowingly acting in ways that would subject others to the risk of significant harm. Because we know so little about the risk, any precautions we can take are much more speculative than in the *WAIS* case. We are significantly more uncertain about the extent to which our precautions actually reduce the risk; and, as we do not know whether or by how much the risk is increasing or decreasing, we cannot judge whether by taking precautions we are not, by diverting resources from other projects, ultimately increasing the risks future people face. None of our actions in asteroid doomsday thus directly involve weighing our interests against those of others or on presuppositions about others and their agency; the degree of uncertainty we face means that we simply lack the relevant information for them to do so.

This suggests that the character of our underlying actions is not only morally relevant to justice under uncertainty but that it also plays an important role in differentiating *WAIS* from *asteroid doomsday*. This important factor is left out of consideration if, as on a universal consequentialist approach, morality focuses only on the consequences of our actions, and the scope of our obligations is determined only by where goodness can be gained or lost. Because it fails to consider these morally relevant aspects of our actions, the consequentialist account can be criticised for being *underinclusive*. Crucially, if the character of our actions is indeed as morally relevant as I argued, this failure means the consequentialist account is normatively inaccurate, in addition to not being sufficiently action-guiding. I earlier defined normative accuracy as the ability to correctly identify the morally relevant features of an action. The consequentialist account is normatively inaccurate as it fails to capture the extent to which our actions are based on presuppositions about future others and the extent to which these presuppositions affect our obligations to them.

The objections discussed in this section raise a set of broader questions about the substance of consequentialism and point to issues which consequentialist theories as a whole may face. Fully discussing these concerns falls outside the scope of this book. Nonetheless, I believe that a good starting point to overcome some of these issues is to amend the underlying account of scope, though it is not my ambition to convince all consequentialist readers of this. For the purpose of climate justice, starting from a different account of scope can help us make important steps towards meeting the climate challenge, irrespective of the substantive theory of justice we later adopt. In the following section, I will begin to formulate such an account by suggesting one way to fix the scope of our obligations in the face of uncertainty. Compared to the consequentialist account just discussed, my proposal provides a more action-guiding and normatively accurate, and therefore more successful, way of setting the scope of our obligations. It therefore provides a more adequate basis on which to ground a substantive theory of climate justice. Later on in this chapter I will discuss whether and how the consequentialist account of scope could be modified in light of my arguments to better respond to the climate challenge.

An alternative approach

In the previous section, I argued that the consequentialist approach to scope pays insufficient attention to the presuppositions about others on which actions imposing foreseeable risks are based. For this reason, it fails to provide a normatively accurate and sufficiently action-guiding account of *whom* we ought to consider for the purpose of climate justice, and of *why* we ought to do so.

An account of scope that also takes account of the character of our actions provides a more successful, in the sense of being action-guiding and normatively accurate, way to deal with the problems uncertainty poses for climate justice. Assessing our actions and the presuppositions about others on which our actions are based allows us to take account of additional morally relevant factors without relying on an exact knowledge of our actions' consequences or their probabilities. These relevant factors are what I have called the foreseeability of a risk, and whether a risk threatens people's most fundamental needs.

In what follows I propose one possible way in which an account of scope can take account of the character of our actions and the underlying presuppositions about others. I argue that risks should only fall within the scope of climate justice if the following two necessary and jointly sufficient conditions are met:

1   The risk is foreseeable to a reasonable agent at the time of acting or deliberating about the need to act, where foreseeability is understood as

   (a) Knowing that one's actions are interfering in the risk in ways that increase its magnitude;
   (b) Having the knowledge and ability to disrupt this interference.

2   The risk threatens people's abilities to access the fundamental necessities they need in order to live autonomous lives.

My arguments are grounded on two considerations. On the one hand, actions on foreseeable risks are necessarily based on presuppositions about the capabilities and vulnerabilities of the risk-exposed, and about how these may be affected by the risk. On the other hand, agents have a duty to respect others by not giving undue weight to their own interests over others' capability to live autonomous lives.

For the purpose of discussion, my arguments can be divided into three parts: first, the foreseeability of a risk, and how it impacts the presuppositions about others which underlie the risk-taking; second, the nature of the interests and goods that are put at risk; and third, the duty to respect others as autonomous agents and how we can meet or violate it when engaging in risk-imposing activity. I will sketch my proposed approach in this section and flesh out the concept of foreseeability in the following section.

Foreseeability

The first strand of my argument is that, whenever we can foresee a risk, any action we take about that risk necessarily involves weighing the interests of one set of agents in imposing the risk against the interests of another set of agents whose safety decreases as they are exposed to greater risks.

I understand "actions about a risk" as actions and omissions intended and consciously performed by the agent or agents. They include taking and not taking precautionary measures, which may involve acting so as to increase or decrease the risk or doing nothing (which in turn may also increase or decrease the risk). I take safety to describe the absence of harm or risks, and hence to decrease if one suffers harm or is exposed to more or greater risks.

Although the set of agents with an interest in the risk-imposition and those exposed to the risk may overlap, in the case of climate change they are for the most part distinct. Moreover, in this book I am only concerned with risks imposed on future people by present people. In the cases I consider, present people are thus both the ones imposing the risks and the ones whose interests are met by the risk-imposition. As future people are the ones exposed to the risk, I will assume them to have an opposing interest in minimising climate risks. Throughout this chapter I will therefore speak of the risk-imposing and risk-exposed agents as distinct from one another. I will also assume that the risk-imposition is in the interest of the risk-imposing agents and contrary to the interests of the risk-exposed.

I earlier gave a rough definition of "foreseeable" to mean two things in this context:

1   The risk-imposing agent knows that her actions are interfering in a risk in ways that increase its magnitude;
2   The agent has the knowledge and ability to disrupt this interference.

These conditions are relevant for establishing duties of justice in situations of uncertainty for the following reasons. That a risk is foreseeable shows that, even if the probability of harm is unknown, the agent can be held accountable for her actions and in turn be held at fault for the risk-imposition. It further makes it the case that any action she takes about the risk in question is based on presuppositions about the risk-exposed and their capabilities.

Imagine the following scenario. Martin enjoys digging holes in the ground to study soil composition around the country and is now considering whether to dig in a patch of publicly accessible land close to his home. To gather enough information, he needs a hole so deep that a fall would cause at least severe and possibly even fatal injury. Martin knows this and knows also that by digging a hole he is increasing the risk that someone may fall into it, even if he does not know by how much. This situation is one of uncertainty, since he knows that there is a risk of harm but is unable to attribute a probability to it. Yet Martin understands that falling into deep holes can be dangerous, he knows his digging significantly increases the risk of danger to passers-by, and he can easily avoid increasing the risk by not digging a hole. Martin's decision on whether to go ahead with the digging or take precautions to minimise the risk of harm therefore inevitably involves weighing his interest in digging against others' interests in not being injured. That is, because the risk is foreseeable, his actions are based on presuppositions about the respective interests involved and their relative weight. Recall that, on the definition of presuppositions I gave, this does not imply that Martin must be aware of making this trade-off; rather, it is a conceptual precondition of an action involving a foreseeable risk.

What matters for the scope of an agent's obligations is that, when the conditions of foreseeability are met, it ought to become known to the agent, and it consequently becomes a feature of their action, that their actions are imposing a risk on others. Regardless of how the activity goes ahead – whether the agent does take increased precautions or not – her action involves weighing the interests of the risk-imposing agent against those of the risk-exposed others. This involves presuppositions about the interests of the risk-imposing and the risk-exposed agents as well as their relative weight. Because actions on foreseeable risks are based on such morally loaded presuppositions, they ought themselves to be the subject of moral evaluation, regardless of whether we can assign probabilities to their possible outcomes.

Autonomy conditions

The second part of my argument holds that foreseeable risk-imposition becomes a matter of justice whenever a risk is imposed on the conditions necessary for agents to live autonomous lives. This sentence contains three distinct claims.

First is the assumption that people need certain fundamental conditions to be fulfilled in order to be able to live as autonomous agents. One way to define these fundamental needs, on which I am going to rely throughout this book, is as the necessary conditions of autonomy or autonomy conditions. Recall that these are the central element of Valentini's autonomy view of justice introduced in Chapter 1 (Valentini 2011, 2013). The conditions necessary to live an autonomous human life plausibly include things like the availability of shelter, nutrition, health care, and freedom of movement (Valentini 2013: 498). Because these conditions are very minimal and fundamental to human life, we can assume they will continue to be important to future generations, including remote future generations.

Second, because these conditions are of such fundamental importance to all agents, it is a matter of justice to protect each person's access to them to the extent necessary to live an autonomous life.[8] Each agent has a claim of justice not to be deprived of the conditions necessary to live an autonomous life (Valentini 2011, 2013).[9]

Third, it follows from this that undermining or imposing a foreseeable risk on people's autonomy conditions also raises questions of justice. Because conditions of autonomy are important enough to give rise to obligations of justice, it equally ought not to be permitted for others to impose unjustified, foreseeable risks on them. Once we establish that agents have a justice-based claim against being deprived of their autonomy conditions, it follows that they should also have a claim against others imposing foreseeable risks on them, though claims against risk-imposition may likely be weaker and more easily outweighed than claims against direct violations of the conditions of autonomy. Note that, beyond adopting her interpretation of our fundamental needs, these three assumptions do not rely on Valentini's autonomy view but are also in line with the recipient-based view of justice.

We are therefore under an obligation to respect each agent's entitlements to the conditions of autonomy whenever we can foresee that our actions will impose a risk of harm on them. Consider Martin again. By digging holes in the ground,

Martin imposes a foreseeable risk on passers-by to fall, get stuck, injure themselves, or even die; in other words, he imposes a risk on their autonomy conditions. Because these persons have a claim of justice against such risk-imposition, Martin has a corresponding obligation to act in ways that respect these claims and fairly balance them against competing claims, for example, to freedom of expression, movement, or learning, which may favour his digging holes.

This leaves open the question of what respect for or the balancing of others' entitlements requires in substance, in other words, which foreseeable risks, or which degree of risk-imposition, may in fact be justified all things considered (see also Valentini 2013: 497). This question is crucial for a theory of intergenerational climate justice but must be answered when specifying the content rather than the scope of such a theory. I will therefore set it aside for the remainder of this book. The important claim for our purpose is the following: if our actions impose a foreseeable risk on others' autonomy conditions, we must extend considerations of justice to the risk-exposed.

Nonetheless, it will be useful to sketch what I take some of the relevant conditions for an autonomous life to be, as these are going to determine which risks fall within the scope of justice and which do not. Formulating a full list of the conditions of autonomy is a task that would require a complete, substantive theory of justice and that falls outside the scope of this book. We can, however, develop an account of scope without a substantive theory of justice and autonomy. The account of scope I am going to develop will be applicable to whichever view of autonomy one holds. What matters for our investigation is that we can assume any plausible account of the conditions of autonomy to include things like life, bodily integrity, freedom of movement, shelter, education, nutrition, and health care. These are universally accepted as fundamental needs and sufficiently uncontested as conditions necessary to live autonomous lives. Importantly, these are all things that have already been affected by the impacts of climate change or that risk being undermined in the future (IPCC 2014a, 2014b, 2018). Thus, no matter which account of autonomy one chooses, as long as it includes this core set of conditions we may say that the impacts of climate change are imposing risks on future people's conditions of autonomy. In the remainder of this book, I am therefore going to assume that the impacts of climate change pose a threat to people's autonomy conditions, either by directly violating them or by imposing foreseeable risks on them.

The duty to respect other as agents

The third and final part of my argument reaffirms our duty to respect future others as agents whenever we engage in actions on foreseeable risks, that is, whenever our actions are based on weighing future persons' interests against our own.

No matter how little we know about specific needs and preferences of future people, we know that insofar as they will be moral persons and agents they are going to be entitled to the basic conditions of autonomy. We, in turn, have an obligation to respect others' entitlements to the conditions of autonomy. These are the basic requirements of justice.

Whenever our activities impose foreseeable risks on future people, they are based on underlying presuppositions about the capabilities and vulnerabilities of these people and on weighing these against our interests in the risk-imposing activity. This is so regardless of whether the risk will eventuate and regardless of whether we can assign a probability to it eventuating. In other words, even if we are uncertain about the consequences of our actions and whether they are ever going to affect the autonomy conditions of future people, the mere fact that our actions impose a foreseeable risk on future persons implies that they are based on certain presuppositions about these others and how they may be affected by our actions.

Recall that these presuppositions of activity need not be acknowledged by the agent. They do, however, form part of our actions. With regard to foreseeably risk-imposing activity this means that regardless of an action's outcome, future persons presupposed by it fall within the scope of that action. It also means that these actions can be more or less respectful towards future entitlements based on how they seem to weigh the presupposed capabilities and vulnerabilities against our interests. Irrespective of their consequences, actions imposing foreseeable risks are thus a vehicle through which we as agents can act in ways that respect others' conditions of autonomy or fail to do so. Because we ought to respect others' entitlements to the conditions of autonomy, engaging in such activity has implications for our obligations of justice. That is, if we engage in actions imposing foreseeable risks we ought to ensure we do so in a way that respects others' conditions of autonomy.

Although we are uncertain about the consequences of our actions, we can respect the future risk-exposed and their entitlements by acknowledging that our actions involve a weighing of interests and by ensuring we weigh our respective interests in ways that give due consideration to future persons' entitlements. Let me give an example to show what this can mean in practice. First, take my previous *WAIS* example again. At the moment we are pursuing a BAU pathway, with CO_2 emissions reaching a record high in 2012 (Crippa et al. 2021). As it imposes foreseeable risks on future persons, our pursuit of a BAU pathway relies on presuppositions about future people. It also portrays a certain level of consideration for their interests: our actions seem to presuppose that present interests in avoiding mitigation costs weigh considerably more than future interests in being free of climate risks and thus able to enjoy the conditions of autonomy. In other words, present generations are imposing a foreseeable risk on future people, but their actions show little consideration for the costs their emissions will impose on future persons in terms of risks to or actual violations of their autonomy conditions. These actions suggest a lack of respect for future agents and their conditions of autonomy.

Consider now a slightly modified scenario:

> *M-BAU. All else being equal, present generations are now providing significantly more funding for research into renewable energy sources and CO_2 removal technologies. Some of the previously fossil fuelled technologies have successfully been replaced by less emitting alternatives, so that global greenhouse gas emissions have begun decreasing. While many of the same emitting*

activities as above are being pursued, the entire climate strategy is notice-
ably more oriented towards maintaining a safe environment for, and hence
protecting the autonomy conditions of, future people.

M-BAU shows greater consideration for future people's entitlements than a BAU scenario. This does not mean that the strategy pursued in this scenario is ideal or sufficient to fulfil the requirements of climate justice. Yet it shows that, even though actions in *M-BAU* still impose foreseeable risks on future people, there are ways in which we can modify our actions and the risks we take to ensure we more fairly balance the entitlements of future persons against our present interests.

Defining what due consideration requires is a question for the substance of justice and as such lies beyond the scope of this book. What matters for the purpose of scope-setting is that whenever our actions impose foreseeable risks on future others and are therefore based on presuppositions about them, we ought to respect these others by extending the basic considerations of justice to them, in other words respecting their entitlements to the conditions of autonomy. Two additional comments about substance seem appropriate at this point. On the one hand, it is very likely that in the case of climate change giving due consideration to the interests of others will require taking significantly more precautions than is currently the case, especially since possible outcomes – such as the total melting of the WAIS – are so harmful.

On the other hand, respecting the risk-exposed will, at least in most cases, likely not require that we fully eliminate a risk. Take, for example, risk-imposition through air travel. Air travel imposes foreseeable risks on persons on the ground.[10] On my proposed account it therefore requires us to fairly consider the security of these persons' autonomy conditions. These persons could rightly complain if, for example, someone was to catapult an unmanned, uncontrollable plane into the air, foreseeing it may strike their nearby village as it comes down, yet not taking any precautions to warn its inhabitants about or shield them from the impact. Fairly considering their interests, however, would plausibly not require fully abstaining from flying any type of aircraft. It seems plausible to say that the interests of the risk-exposed are respected if, as is the case in modern air travel, both airplanes and their travel are regulated, and meet stringent safety standards, and aircrafts are only allowed to start and land in designated areas. Such precautions would seem to adequately balance the interests of the risk-imposing agents with those of the risk-exposed persons.

Let me now summarise what the discussion so far implies for the scope of climate justice. I argued that even if we cannot foresee the exact harm our actions will cause because we are acting in a situation of uncertainty, if we can foresee that we are imposing the risk of harm on others, then our risk-imposing actions involve presuppositions about the risk-exposed, more specifically about our respective interests in imposing or being free from risks and these interests' relative weight. If our actions impose a risk on others' conditions of autonomy, we are imposing a risk on what the risk-exposed need to live as autonomous agents, thus on fundamental needs secured by justice. Whenever our actions impose foreseeable risks on others' – including future others' – conditions of autonomy, we ought to respect these persons' basic entitlements of justice by giving due consideration to their interests in the process. This applies to actions related to climate justice, too. Most activity relevant to climate justice, including continuing on a BAU trajectory or taking measures to mitigate future climate changes, imposes or minimises certain risks on future people. In these cases, too, we have an obligation to extend the basic requirements of justice to the risk-exposed future people; in other words, we ought to include the risk-exposed within the scope of climate justice.

Understanding foreseeability

The concept of foreseeability is crucial to my approach to justice under uncertainty. In this section, I will unpack the concept and flesh out its relevance to my argument. I will answer two important questions: why is foreseeability relevant to whether a risk gives rise to duties of justice? And what does it take for a risk to be foreseeable to an agent?

Let me begin by tackling the first question. The notion of foreseeability is necessary because it allows us to hold the agent at fault for risk-imposing activity despite any uncertainty about the actions' effects. On the understanding of duties under uncertainty I propose in this chapter, that the agent be able to foresee the risks imposed by the activity under consideration is a necessary condition for justice. I argued that in situations of uncertainty, duties of justice are grounded on two premises: first, that the action under consideration be based on presuppositions about the risk-exposed agents and how the risk may affect their capacities to live autonomous lives; and, second, that agents have a duty to respect others' entitlements to the conditions of autonomy.

I also argued that what matters morally is not only the outcome of our actions but whether we adequately respect the risk-exposed others as we perform our activities. This includes giving due consideration to their interests in the conditions of autonomy whenever we perform actions that are based on an underlying weighing of our respective interests. Only if we can foresee the risks imposed by our actions can we be responsible for how our actions may affect the risk-exposed and be required to respect their agency as we engage in these activities. Foreseeability allows us to establish whether we can be held at fault for and thus under a duty of justice with regard to the risk-imposing action.

Consider, first, the following scenario:

> PEANUTS. Lena prepared a cake for a bake sale to be held at her university. Although she knows that a number of students and staff are allergic to peanuts, she decided to add peanuts to her cake. She could easily have swapped them for almonds but chose not to, nor did she inform customers that the cake contains peanuts.

Clearly, Lena is at fault for imposing the risk of an allergic reaction on staff and students as well as for any actually resulting harm. It was clearly foreseeable to her that she was imposing this risk on others, and she could easily and at no excessive cost to herself have eliminated the risk. That she nevertheless chose to impose the risk on others implies that she is at fault for the risk-imposition.

Let's now modify the scenario to investigate what exactly is doing the work in grounding Lena's fault for the risk-imposition. Consider a second scenario:

> *FLOUR. Some staff and students have developed a rare but severe allergy to the type of flour used in Lena's cake. This is a very rare allergy known only to a few medical researchers and which even the sufferers are unaware of. Lena sells her cake without informing her customers of the flour it contains.*

By selling her cake without an appropriate notice, Lena is imposing a risk of harm on her customers. Yet in this case, it is less clear that she can be held at fault for doing so. It is unreasonable to expect Lena to be aware of a medical condition this rare, nor could she have gained any knowledge of it by asking staff and students for known allergies prior to baking, as they themselves are unaware of living with this allergy. Because she could not possibly have foreseen the risk-imposition, she cannot plausibly be held at fault for either the risk-imposition or any actually occurring harm.

Flour suggests we may not hold an agent at fault for a risk-imposition she could not have foreseen (and therefore could not have avoided or minimised either). Imagine, however, a slightly modified version of this thought experiment:

> *M-FLOUR. Lena was told about the allergy and its prevalence at her university and was therefore aware of the risk that using this type of flour would impose on others. Yet when baking she grabbed the first flour packet from her cupboard, quickly skimmed over the ingredient list and failed to realise she was about to use the very type of flour she had been asked to avoid.*

By selling her cake, Lena is imposing a risk on others that she does not foresee at the moment of acting, yet in this case it is far from clear that she cannot be held at fault for the risk-imposition. In fact, it seems she can and should be held at fault for imposing a significant risk on others, for she really should have known what type of flour she was baking with. In other words, Lena *should have foreseen the risk*. The requirement for foreseeability has to be controlled for negligence: under all plausible definitions of reasonableness, it can reasonably be expected of Lena that she know which ingredients went into her cake – provided all were correctly labelled and Lena was able to read – especially given that she was aware of the allergy and its severity. In this case, obtaining the information about the possible consequences of her actions would not have been too onerous either. *M-flour* suggests that, if the agent's failure to foresee a risk she imposes on others is due to the agent's own negligence, she can nevertheless be held at fault for the risk-imposition through her actions.

The examples so far suggest that in order for an agent to be at fault for a risk-imposition, it must be reasonable for her to foresee the possible effects of her actions. A final example can show whether foreseeability suffices to establish the agent's fault in the risk-imposition.

> *TRIPPING. Lena knows about the flour allergy. She plans to use an alternative flour, yet as she is getting ready to bake and moves to place a packet of the dangerous flour back in the cupboard she trips over, through no fault of her own, spilling some flour onto a colleague who happens to be standing nearby. He immediately develops a severe allergic reaction to it and has to be rushed to hospital.*

Here, it is not clear that Lena is at fault for the risk imposed on her colleague by her spilling the flour. While we can say that her actions imposed a risk on and harmed her colleague, this alone is not enough to establish that Lena has violated her duty to respect his autonomy conditions. Although she foresees the risk-imposition, she cannot control it. The fact that she cannot control the risk – that is, reduce or remove it – seems to make a relevant difference to whether she can be held at fault and subject to judgements of justice for imposing it on him or for it eventuating. Because uncontrollable actions are not an appropriate representation of the agent's agency, we cannot use them and the underlying presuppositions to determine what justice requires of that agent. Thus, if an agent cannot control her actions, she cannot be subject to judgements of justice for their effects on others. This indicates that narrow foreseeability alone does not establish whether an agent can be held at fault for a risk-imposition but that it must be complemented by the requirement that the agent be able to control her actions.

The discussion so far suggests that agents can be subject to judgements of justice for their actions if they know, or should have known, about the possible effects of their actions and if they are (or were at a previous point in time) able to control their actions.[11] To meet these requirements, the wider concept of foreseeability I propose in this chapter includes a narrower foreseeability condition and a control condition. In response to the second question I posed at the start of this section we can now say that foreseeability is defined as follows.

FORESEEABILITY:

1   *The risk-imposing agent knows or can reasonably be expected to know that her actions are interfering in a risk in ways that increase its magnitude;*
2   *The agent has or can reasonably be expected to have the knowledge and ability to disrupt this interference.*

This definition is applicable to situations of uncertainty. Even if we cannot assign exact probabilities to the consequences of our actions, if we know that our actions are such that they amplify a risk then we know, first, that we are imposing a risk on others; and, second, if we know the recurrence of the risk-imposing actions, we are able to get a sense of the likely magnitude of the risk.

It is important to emphasise that this notion of foreseeability does not excuse negligence. An agent acts negligently if she is not aware of the possible harmful consequences of her actions when she could and should have been (Ekeli 2004: 424–425). To avoid acting negligently, the agent must make reasonable efforts to obtain information about the possible consequences of her actions, whether her actions are likely to impose risks on others and, if so, how she could reduce or avoid these risks. How to define reasonable efforts is a controversial but clearly very important question which a substantive theory of justice, including climate justice, will have to answer. For the purpose of scope-setting, it can be momentarily set aside.

Nevertheless, a few comments on how the notion of reasonable efforts relates to the rest of the proposed approach are in order. Because the understanding of justice defended in this book centres on the protection of people's conditions of autonomy, so will the concept of reasonable efforts. In turn, what reasonableness requires will largely depend on the account of autonomy one supports. For one, whether an effort counts as reasonable should be determined in reference to and proportionally to the costs it imposes on the agent's conditions of autonomy. For another, the conditions of autonomy mark an upper boundary of what can possibly be demanded of an agent, so that agents can only be required to seek out information about their action's effects as long as the costs of doing so do not endanger their autonomy conditions.

As I am not working with any specific account of autonomy I do not have access to a full account of reasonableness either. Nevertheless, the relation between reasonableness and autonomy just fleshed out allows us to draw some approximate comparisons between risks. We have, for example, been able to acquire information about the impacts of anthropogenic emissions on the climate and possible mitigation and adaptation measures relatively easily and without imposing excessive costs on present generations. We have less information on the risk of asteroid impact, especially in terms of causation and whether our actions are in any way interacting with the risk. Compared to climate research, investigating strategies to avoid possible asteroid impacts, how likely these are to function correctly, and whether, if they do, deploying them will actually lower the risk of impact would plausibly be much more costly. Similarly, the efforts required of Lena to inform herself about an obscure allergy are significantly greater than the cost to her of carefully reading through a list of ingredients.

When applying the notion of reasonableness as a decision-making tool it is important to keep in mind that – because of its connection to the conditions of autonomy – its requirements will be agent-relative. That is, it is going to require different efforts depending on the agent in question, that agents' capacities, and the costs *to that agent* of acquiring the relevant information. With regard to climate justice, for instance, a layperson without a degree in the natural sciences cannot reasonably be expected to understand future climate changes to the same degree as a specialised government body or a research institution.[12]

In this book I am only concerned with the duties of organised collective agents with regard to climate justice, which simplifies the task of approximating a notion of reasonableness that we can work with. We only need to get a grasp of what can reasonably be expected of larger organisations, companies, and public bodies, which are either themselves authorities on climate change, include departments specialised on the topic, or have the resources to gain access to the most current climate science. We can, at the very least, reasonably expect these agents to be aware of the latest scientific research on climate change and its risks as well as to understand scientists' main conclusions and predictions.

With these clarifications in mind, we can now return to our definition of foreseeability. Condition *1* of our definition, with addition of the requirements to make reasonable efforts and not act negligently, establishes whether it is foreseeable to the agent that her actions are imposing a risk on others. Yet even if the risk is foreseeable, we cannot hold the agent at fault for a risk-imposition if she was at no point able to control her actions and thus unable to refrain from or limit the risk-imposition. If, for example, Anne suffers from uncontrollable muscle spasms that are causing her to hit Bill, she cannot be subject to judgements of justice for the harm she is inflicting on him. The definition of control must thereby also include the agent's past ability to exert control over her actions. This is so that agents cannot use past but has freely entered into commitments to excuse present wrongdoing. If, for example, Anne had freely and under no threat to her own autonomy conditions taken a pill she knew would cause uncontrollable muscle spasms and make it more likely she would hit Bill, it seems she can at least partly be held at fault for hitting him, even if the hitting itself was not under her control.[13]

Condition *2* of foreseeability thus builds on condition *1*, establishing whether the agent has the necessary level of control over what she can foresee are risky activities. If the agent is unable to control her actions, she may not be subject to judgements of justice for the risks they impose on others. That is, the presuppositions on which her actions are based cannot be used to determine whether or not the agent acquires obligations of justice.

As a final point, it is worth noting that this notion of foreseeability may positively affect the presence of ignorance and "unknown unknowns" that still mar our ability to predict future climate changes. Because it imposes an obligation on agents, including agents with significant capacities for research, to make reasonable efforts to understand the possible consequences of their actions, this concept of foreseeability can promote further research into the effects our actions have on the climate. Insofar as it does, it can help improve our general understanding of climate change and may contribute, if only to a small degree, to eliminating some amount of ignorance about its future impacts.

Consequentialism revisited

The concept of foreseeability proposed in this chapter enables an account of scope to differentiate between those risks that give rise to duties of justice and those that do not. One of the objections I raised against the actual universal consequentialist account of scope is that it fails at making precisely this distinction, in other words that it is overinclusive. Given what I have just argued, one obvious response to my objection would be to amend the consequentialist account to include a condition of foreseeability. In this section I will argue, first, that while this would solve the problem of overinclusiveness the account would then cease to be a consequentialist account; and, second, that even if one were to make this amendment, the account would still face a problem of underinclusiveness.

Amending the consequentialist account to include the criterion of foreseeability would allow it to distinguish between what I argued are justice-relevant and justice-irrelevant risks. Rather than viewing all risks as possible sources of duties of justice, on the amended consequentialist account one would only consider those actions as giving rise to duties of justice that meet the conditions of foreseeability. That is, one would only consider actions the agent knows are interfering with a risk in ways that increase its magnitude and from which the agent can fully or partially abstain. This modification allows the consequentialist account to draw what I argued is a crucial distinction between the *asteroid doomsday* and *WAIS* risks. Earlier in this chapter I suggested that, contrary to *WAIS*, in *asteroid doomsday* we have no evidence showing that we are acting in ways that increase the risk of impact and that we are therefore unable to get even a sense of the magnitude of risk and whether our precautions would decrease it. With regard to justice, I take this to mean that it would not be reasonable to hold any set of people to be subject to obligations of justice to reduce this risk. By incorporating the notion of foreseeability, the consequentialist account would fall in line with this intuition and only consider *WAIS* as possibly giving rise to duties of justice.

The amended account seems, however, to amount to an abandonment of consequentialism. Recall that, on the consequentialist theory under consideration, "moral rightness depends on the consequences for all people or sentient beings" (Sinnott-Armstrong 2019). Adopting the suggested condition of foreseeability introduces limiting factors other than an action's consequences as a way to determine an agent's obligations, thus departing from the sole pursuit of impartially considered good effects. This change in character can be argued to call into question the overall coherence of the consequentialist theory, but it need not be a reason to abandon the account altogether, though the new approach may demand accurate justification.

Yet still, if one agrees with what I have previously argued in this chapter, even the modified consequentialist account remains underinclusive with regard to what it takes to be morally relevant. Including a condition of foreseeability improves the account by limiting the risks it considers to only those that are foreseeable, yet the moral rightness of these actions, and hence whether and why they do in fact give rise to obligations, continues to be determined solely by their consequences on sentient beings. This is problematic in situations of uncertainty, as our extreme cluelessness about the consequences of our actions is going to severely limit our ability to morally evaluate our actions. It is even more concerning if one agrees that there may be additional factors, beyond an action's consequences, that are relevant to its moral evaluation under conditions of uncertainty. Unless one firmly believes that only the consequences of our actions matter, it is plausible, as I have argued, that it be the presuppositions on which our actions are based which determine whether or not our actions give rise to obligations of justice. Insofar as one agrees that the character of our actions is in this way relevant to their moral evaluation, the consequentialist account can be charged with being normatively inaccurate for its failure to consider it.

This is problematic in itself if one takes seriously the charge of normative inaccuracy. At the same time, a failure to consider the character of our actions can be cause of concern when it comes to the substance of our obligations. An account that takes only the consequences of our actions as the source of obligations is likely, for example, to overlook the different sets of persons about whom we make presuppositions and who are variously affected by our actions, while instead working to maximise the overall impartial good brought about by our actions. This speaks to familiar and fundamental differences between consequentialist and deontological theories, the details of which go beyond the scope of this book. I only note that the charges of underinclusiveness and normative inaccuracy are relevant to the question of scope but have important ramifications beyond that, too.

In contrast to the consequentialist account, the approach I proposed in this chapter can be both action-guiding and normatively accurate in situations of uncertainty. By pointing out a morally relevant difference between foreseeable and unforeseeable risks, it provides agents with practical guidance on which category of risks to focus on in situations of multiple uncertainties.[14] And because it extends beyond the consequences of our actions to also consider their character, it fares better than the consequentialist account in identifying the morally relevant features of risk-taking in situations of uncertainty. It is therefore better suited to meet the intergenerational climate challenge.

Conclusion

In this chapter, I introduced the first criterion for a successful account of the scope of climate justice. I argued that, in order to successfully meet the climate challenge, an account of the scope of climate justice must be able to deal with the uncertainties of our climate predictions.

I outlined the types of risks and uncertainties we face when predicting future climate changes, arguing that the situation is best characterised as one of uncertainty rather than pure risk. This is because, in most cases, we are unable to assign a sufficiently accurate probability to possible future outcomes.

I then argued that, given our level of uncertainty, a universal consequentialist account of scope is inadequate as an account of the scope of climate justice. Because it only focuses on the consequences of our actions and only sees people as bearers of utility, it can be criticised for being both overinclusive and underinclusive. On the one hand, it is overinclusive because it cannot, without further qualifications, differentiate between types of risks that intuitively call for a different moral response. It therefore fails to be action-guiding in situations of multiple uncertainties. On the other hand, it is underinclusive because it fails to take account of the character of our actions, even though, as I also argued, this is morally relevant to an account of intergenerational justice. As an account it therefore fails to be normatively accurate.

I argued that an account of the scope of justice can respond more successfully to uncertainty if it takes our actions, especially the presuppositions about others on which they are based, as the subject of moral evaluation and grounds of our duties of justice. On the one hand, this alternative approach achieves more action-guiding results by differentiating between categories of risks based on a specific notion of foreseeability, which uses a narrow condition of foreseeability and a control condition to determine whether an agent can be subject to obligations of justice with regard to her actions. On the other hand, this way of thinking about the scope of justice is normatively more accurate than its consequentialist counterpart as it captures the moral relevance of our presuppositions, as well as the importance of respecting other agents and their autonomy conditions as we act.

In the next chapter, I will defend the second criterion that an account of scope the scope of climate justice must meet, C2, arguing that an account of scope must be able to respond to climate change as a complex problem of justice.

Notes

1 Climate change is already having harmful effects in the present, such as the melting of ice sheets and sea-level rise (Mouginot et al. 2019), yet its impacts are predicted to become more severe and harmful over time (IPCC 2014a, 2021). In this book, I will focus on the harmful effects that climate change is predicted to have on future people due to greenhouse gas emissions before their time.

2 For a more detailed account of the complexity and scientific uncertainty involved in climate modelling, see Oppenheimer et al. 2014; Tebaldi et al. 2008.

3 Much of the uncertainty that pervades even assigned probabilities comes from genuine underlying uncertainties, such as the possibility of feedbacks from the climate system, or the role that socio-economic systems and their ability to adapt and respond to the climate play in determining the harmfulness of a certain impact (for more detail, see Hartzell-Nichols 2017, especially chapter 4).

4 This book focuses on the human element of climate justice. I am therefore bracketing other sentient beings and limiting my analysis to persons.

5 In a recent reply to Greaves, Yim suggests that most if not all actions, including supposedly simple ones, are in fact cases of complex cluelessness. The reason is that all other-affecting actions cause chains of systematic impacts on others, along the line of which is likely to be an impact of complex nature (Yim 2019). This view further strengthens the case for evaluating actions related to climate change as cases of complex cluelessness.

6 The first condition is adapted from Henry Shue (2010, 2015). He uses his version of the condition to establish "threshold likelihood." This is itself one of three conditions – together with the threat of massive loss and non-excessive costs of precaution – that establish the need to take precautions despite low or unknown probability that the harmful event will occur.

7 A BAU option is not in the interest of all present people, especially younger present people and those who are already being affected by the impacts of climate change. However, for the purpose of this book I assume that avoiding mitigations costs by pursuing a BAU option is still in the interest of the majority of present agents relevant to climate justice this investigation is concerned with.

8 This is only the most basic principle of justice and leaves open important questions such as how to distribute a resource surplus or what to do if there are insufficient resources to allow everyone to enjoy sufficient autonomy conditions. These are questions that a complete theory of justice would have to answer but which can be bracketed for the sake of this investigation.

9 Valentini's autonomy view holds that in order to respect others as autonomous agents – and provided we have the ability to refrain from undermining their autonomy

conditions – we have both a negative duty not to undermine their autonomy conditions and a positive duty to help them fulfil their autonomy conditions if we can do so without excessive sacrifice (Valentini 2013: 497). While my own view of justice draws on the autonomy view, for the purpose of scope-setting we also do not need to accept her claims on the substantive content of justice. For the purpose of my argument, it is enough to assume a negative duty not to undermine others' autonomy conditions. There is no need to settle whether we also have a positive duty to help others fulfil their autonomy conditions.

10 As well as on persons in the aircraft, but I am assuming these persons are willingly taking those risks.

11 I take much of this definition from Valentini's (2013) understanding of justice, which she links to the involvement of the agent's moral agency through a foreseeability and control condition.

12 For an in-depth discussion of the extent to which a layperson can be expected to understand and assess scientific testimony, see Anderson 2011. In addition, there are ways in which individuals can be prevented from understanding or accessing the relevant information which are themselves injustices. For an in-depth treatment of these epistemic injustices, see Fricker 2007; for a view on how social practices and institutions can impede the formation of true beliefs, see Buchanan 2002.

13 This example expands on a similar example in Valentini 2013.

14 Of course, clashes can still arise between risks that fall into the same category of foreseeable risks; that is, situations can arise in which, due to the scarcity of resources, not all duties of justice can be equally met. When moving beyond questions of scope, a full theory of climate justice would have to specify additional principles or decision-making procedures to deal with such cases.

Reference list

Anderson, Elizabeth. 2011. 'Democracy, Public Policy, and Lay Assessments of Scientific Testimony'. *Episteme* 8 (2): 144–164. https://doi.org/10.3366/epi.2011.0013.

Bamber, Jonathan L., Michael Oppenheimer, Robert E. Kopp, Willy P. Aspinall, and Roger M. Cooke. 2019. 'Ice Sheet Contributions to Future Sea-Level Rise from Structured Expert Judgment'. *Proceedings of the National Academy of Sciences* 116 (23): 11195–11200. https://doi.org/10.1073/pnas.1817205116.

Buchanan, Allen. 2002. 'Social Moral Epistemology'. *Social Philosophy and Policy* 19 (2): 126–152. https://doi.org/10.1017/S0265052502192065.

Crippa, M., D. Guizzardi, E. Solazzo, M. Muntean, E. Schaaf, F. Monforti-Ferrario, M. Banja, et al. 2021. *GHG Emissions of All World: 2021 Report*. Luxembourg: Publications Office of the European Union. https://data.europa.eu/doi/10.2760/173513.

Ekeli, Kristian Skagen. 2004. 'Environmental Risks, Uncertainty and Intergenerational Ethics'. *Environmental Values* 13 (4): 421–448. https://doi.org/10.3197/0963271042772578.

Fricker, Miranda. 2007. *Epistemic Injustice*. Oxford: Oxford University Press. www.oxfordscholarship.com/view/10.1093/acprof:oso/9780198237907.001.0001/acprof-9780198237907.

Gardiner, Stephen M. 2011. *A Perfect Moral Storm*. Oxford: Oxford University Press. https://doi.org/10.1093/acprof:oso/9780195379440.001.0001.

Greaves, Hilary. 2016. 'Cluelessness'. *Proceedings of the Aristotelian Society* 116 (3): 311–339. https://doi.org/10.1093/arisoc/aow018.

Green, Fergus. 2015. *Nationally Self-Interested Climate Change Mitigation: A Unified Conceptual Framework*. GRI Working Papers 199. Grantham Research Institute on Climate Change and the Environment. https://ideas.repec.org/p/lsg/lsgwps/wp199.html.

Guy, Jack. 2019. 'Instability Spreading in West Antarctic Ice Sheet'. *CNN*. 16 May. https://edition.cnn.com/2019/05/16/europe/antarctic-glacier-melt-scli-intl-scn/index.html.

Hartzell-Nichols, Lauren. 2017. *A Climate of Risk: Precautionary Principles, Catastrophes, and Climate Change*. Environmental Politics/Routledge Research in Environmental Politics 26. New York: Routledge.

Hooker, Brad. 2002. *Ideal Code, Real World*. Oxford: Oxford University Press. https://doi.org/10.1093/0199256578.001.0001.

IPCC. 2013. *Climate Change 2013: The Physical Science Basis. Working Group I Contribution to the Fifth Assessment Report of the Intergovernmental Panel on Climate Change*. New York: IPCC.

———. 2014a. *Climate Change 2014: Impacts, Adaptation, and Vulnerability. Working Group II Contribution to the Fifth Assessment Report of the Intergovernmental Panel on Climate Change*. New York: IPCC.

———. 2014b. *Climate Change 2014: Mitigation of Climate Change. Contribution of Working Group III to the Fifth Assessment Report of the Intergovernmental Panel on Climate Change*. New York: IPCC.

———. 2014c. *Climate Change 2014: Synthesis Report. Contribution of Working Groups I, II and III to the Fifth Assessment Report of the Intergovernmental Panel on Climate Change*. New York: IPCC.

———. 2018. *IPCC Special Report Global Warming of 1.5°C*. Incheon, Republic of Korea: IPCC.

———. 2021. *Climate Change 2021: The Physical Science Basis. Contribution of Working Group I to the Sixth Assessment Report of the Intergivernmental Panel on Climate Change*. Cambridge: IPCC.

Knight, Frank Hyneman. 2002. *Risk, Uncertainty and Profit*. Washington, DC: Beard Books.

Lenman, James. 2000. 'Consequentialism and Cluelessness'. *Philosophy and Public Affairs* 29 (4): 342–370. https://doi.org/10.1111/j.1088-4963.2000.00342.x.

Mouginot, Jérémie, Eric Rignot, Anders A. Bjørk, Michiel van den Broeke, Romain Millan, Mathieu Morlighem, Brice Noël, Bernd Scheuchl, and Michael Wood. 2019. 'Forty-Six Years of Greenland Ice Sheet Mass Balance from 1972 to 2018'. *Proceedings of the National Academy of Sciences*, April, 201904242. https://doi.org/10.1073/pnas.1904242116.

Oppenheimer, M., M. Campos, R. Warren, J. Birkmann, G. Luber, B. O'Neill, and K. Takahashi. 2014. '2014: Emergent Risks and Key Vulnerabilities'. In *Climate Change 2014: Impacts, Adaptation, and Vulnerability. Part A: Global and Sectoral Aspects. Contribution of Working Group II to the Fifth Assessment Report of the Intergovernmental Panel on Climate Change*, edited by C.B. Field, V.R. Barros, D.J. Dokken, K.J. Mach, M.D. Mastrandrea, T.E. Bilir, M. Chatterjee, et al., 1039–1099. Cambridge: Cambridge University Press.

Shankman, Sabrina. 2019. 'Coasts Should Plan for 6.5 Feet Sea Level Rise by 2100 as Precaution, Experts Say'. *Inside Climate News*. 21 May. https://insideclimatenews.org/news/21052019/antarctica-greenland-ice-sheet-melting-sea-level-rise-risk-climate-change-polar-scientists.

Shue, Henry. 2010. 'Deadly Delays, Saving Opportunities: Creating a More Dangerous World?' In *Climate Ethics: Essential Readings*, edited by Stephen M. Gardiner. Oxford: Oxford University Press.

———. 2015. 'Uncertainty as the Reason for Action: Last Opportunity and Future Climate Disaster'. *Global Justice: Theory, Practice, Rhetoric*. https://doi.org/10.21248/gjn.8.2.89.

Sinnott-Armstrong, Walter. 2019. 'Consequentialism'. In *The Stanford Encyclopedia of Philosophy*, edited by Edward N. Zalta. Stanford: Metaphysics Research Lab, Stanford University. Summer.

Stern, Nicholas, ed. 2007. *The Economics of Climate Change: The Stern Review*. Cambridge: Cambridge University Press.

Tebaldi, Claudia, Gavin Schmidt, James Murphy, and Leonard A. Smith. 2008. 'The Uncertainty in Climate Modeling'. *Bulletin of the Atomic Scientists*. 22 April. https://thebulletin.org/roundtable/the-uncertainty-in-climate-modeling/.

Valentini, Laura. 2011. *Justice in a Globalized World: A Normative Framework*. Oxford: Oxford University Press.

———. 2013. 'Justice, Charity, and Disaster Relief: What, If Anything, Is Owed to Haiti, Japan, and New Zealand?' *American Journal of Political Science* 57 (2): 491–503. https://doi.org/10.1111/j.1540-5907.2012.00622.x.

Yim, Lok Lam. 2019. 'The Cluelessness Objection Revisited'. *Proceedings of the Aristotelian Society* 119 (3): 321–324. https://doi.org/10.1093/arisoc/aoz016.

3 Climate change as a complex problem of justice

The problem of complexity

While climate change certainly confronts us with individually tough challenges – such as the problem of uncertainty – it is their combination which makes it a particularly complex moral problem. The complexity created by the convergence of these individual issues is an additional hurdle that our moral theories must be able to deal with. Most importantly for this investigation, the complexity of climate change also challenges the ability of our moral theories to set the scope of climate justice. In this chapter I am going to argue that, as a second criterion, an account of the scope of climate justice must be able to respond to climate change as a complex problem of justice. An account of scope must meet the following criterion C2:

> *DEALING WITH COMPLEXITY. An account of the scope of climate justice must be able to respond to climate change as a complex problem of justice.*

It has often been argued that our current moral theories are unable to deal with large-scale environmental problems. Two theorists in particular – Dale Jamieson and Stephen Gardiner – have pressed this concern with regard to climate change. Despite their similarities, their accounts single out diverging causes for our theories' inability to adequately respond to the climate problem. A brief look at their arguments will help clarify the impacts complexity has on climate justice and what it may take to overcome them.

Jamieson lays out the somewhat more dire account, arguing that the root of the problem lies in our lack of appropriate values and concepts suited for dealing with climate change (Jamieson 1992, 2010, 2013, 2014). He argues that climate change poses a uniquely complex challenge, for it combines individual issues in a way that exhibits some marks of a paradigmatic moral problem but not others. These issues include the spatial and temporal reach of climate change, its systemic nature, the complexity of the climate system, and the fact that climate change is world-constituting, that is, that it will repopulate the world by affecting who will be born (Jamieson 2014). However, because our current values and moral concepts evolved in low-population, low-technology societies, Jamieson argues they

DOI: 10.4324/9781003258902-3

are unable to deal with global, intergenerational problems like this one (Jamieson 1992, 2014).

The problem for climate justice, in Jamieson's view, is that our lack of appropriate concepts limits our ability to apportion responsibility for climate change, and thus to assign corresponding duties, on two fronts. On the one hand, we cannot easily apply existing concepts of individual responsibility to the problem. These concepts evolved to work best with a paradigmatic notion of harm and presuppose harms and causes that are individual, readily identifiable, and local in space and time (Jamieson 1992: 148). They allow us to clearly apportion responsibility whenever one individual intentionally harms another, both the individuals and the harm are clearly identifiable, and all are closely related in time and space – if, for example, I was to steal your bike (Jamieson 2013: 39–40). Unfortunately, they do not work as well when applied to large-scale, long-term, and diffuse moral problems like climate change. Contrary to paradigmatic cases of responsibility, climate harms are presumably unintentional, neither the harms nor the responsible agents are clearly identifiable, and both are dispersed across time and space.

On the other hand, Jamieson identifies a second, parallel problem in that climate change also strongly deviates from traditional, statist models of global justice. That is, we cannot clearly apportion national responsibility as we would, for instance, in cases of military aggression. Unlike military invasions, climate harms are a by-product of our actions rather than an intentional infliction of damages, and their impacts and contributions are diffuse rather than focused on one state. Although Jamieson does agree that climate change raises questions of global justice, he thinks the extent to which we can use existing concepts to hold states politically responsible is extremely limited. Instead, rather than in terms of states, it seems more appropriate to frame climate change in terms of major contributors and vulnerable individuals spread across the globe (Jamieson 2010, 2014: 8).

Jamieson believes we need a radical solution in order to solve this problem of fit between our morality and the issue we are facing. We will not be able to repurpose our existing values and moral concepts to successfully apportion responsibility for climate change, he says, but ought instead to formulate and implement new values, moral understandings, and concepts of responsibility. That is, we ought to reformulate our framework of morality to be more in line with the complex new challenges posed by a globalised and technologically advanced world. New values, so Jamieson, will better reflect the interconnectedness of our societies and more adequately match the scale and complexity of the problem of climate change (Jamieson 1992, 2010, 2014).

Jamieson's arguments draw on a broader tradition of environmental thinking which underlines the imperfect fit between our moral theories and complex, large-scale environmental problems more generally. The Deep Ecology movement began raising concerns about our inability to respond to global environmental issues in the 1970s. According to Arne Næss, one of the founders of the movement, the root of the problem was that citizens were not living up to their capacities for identifying with the natural world. He believed this failure had to be addressed by a change within morality, and he sought to shift individual behaviour and policies through new philosophies emphasising the place of humans within the natural world ("ecologically inspired total views") (Glasser 2011; Næss 1973).

Næss was not the only one pushing for change, with others like Ornstein and Ehrlich arguing that the disconnect between the natural world and unsustainable human behaviours could be solved by developing new capacities for perception and judgement (Ornstein and Ehrlich 1989, as cited in Glasser 2011). Similarly, some years earlier Lynn White had argued for the need to replace anachronistic, anthropocentric values with non-anthropocentric, life-affirming axioms (White 1967, as cited in Glasser 2011). These earlier arguments are very close to Jamieson's contention that the complexity problem of large-scale environmental concerns is best solved by moving to new, less anthropocentric moral values and concepts. More importantly, the work of these earlier theorists shows that the problem of theoretical inadequacy for dealing with complexity is an enduring and deep-seated one within environmental philosophy.

While Gardiner picks up on the issue of complexity, too, he offers a more moderate view of the challenge and a more optimistic take on our ability to rescue existing norms and concepts. Recall that as part of his perfect storm analogy, introduced in Chapter 1, Gardiner identifies our current lack of theories able to deal with a problem as complex as climate change as what is causing the theoretical storm. But he does not call for equally sweeping changes to our morality in order to solve the issue. Instead, he hypothesises that the reason we are failing to adequately act on climate change is not an *outright lack* of moral concepts but merely an *inability to access* the relevant norms of climate ethics (Gardiner 2011a, 2011b).

Like Jamieson, Gardiner explains this lack of access in reference to our moral and political theories, which he, too, believes are inadequate for dealing with the complexity of climate change. Yet rather than beyond repair, he believes they are merely underdeveloped in areas that are especially relevant to a moral response to climate change – such as intergenerational ethics, international justice, and scientific uncertainty. Because we lack access to moral norms that can guide our behaviour in these crucial areas, we are unable to adequately respond to, for example, the global or temporal dimension of climate change. This inability, in turn, is what causes the global and intergenerational storms.

In what ways, then, does the inadequacy of our moral theories limit our ability to act justly in a changing climate? Gardiner identifies two main impacts. On the one hand, he argues that the absence of easily identifiable norms makes us vulnerable to the kind of moral corruption discussed in Chapter 1. That is, it makes us more likely to fill the void by distorting the norms we do have in ways that support our own preferences over competing interests – for example, those of future generations. On the other hand, he believes that we cannot – at least not easily – draw on our existing theoretical repertoire to solve the distinct problems of climate change. Consider, for example, the difference between standard intragenerational collective action problems and the intergenerational collective action problem (IGP) which is at the centre of climate change. In intragenerational cases, parties can interact with each other and encourage cooperation by appealing to reciprocal advantages. And because all parties stand to benefit from cooperation, it is collectively rational for them to cooperate with each other. IGPs, on the other hand, where parties are non-overlapping generations who cannot interact with one another, seem much more difficult to resolve by drawing on familiar moral theories. Traditional understandings of reciprocity and cooperation, for example, do not provide for the possibility of cooperation between parties who never exist at the same time. Because earlier generations cannot be benefitted by later ones, it will never be collectively rational for the first generation to cooperate with its successors. The IGP exemplifies how the inadequacy of our theories inhibits our ability to act morally. According to Gardiner, our accounts of intergenerational justice are simply not robust enough to resolve it, and because of a lack in moral guidance, when its turn comes each generation is tempted to pursue its self-interest and engage in morally reprehensible behaviours like intergenerational buck-passing (Gardiner 2003, 2011b: chapters 5 and 6).

Jamieson's and Gardiner's arguments in particular underscore the broader importance of my proposed criterion – that is, of finding a way to respond to the complexity of climate change when thinking of climate justice – while also helping us understand the impact complexity has on a theory of climate justice. They emphasise the limitations which the complexity of climate change imposes on our ability to rely on current moral theories to guide action in the climate emergency. The solutions they propose, however, differ significantly from one another. Whether we will need to develop our theories in relevant areas or formulate new moral concepts altogether is going to make a considerable difference to the account of climate justice or, in our case, of the scope of climate justice, we ultimately formulate. One of the aims of this chapter will be to show, contra Jamieson, that we do not need to devise new moral concepts but can formulate a successful account of scope by developing and repurposing existing values and ideas.

It is just as important for an account of scope as it is for a full theory of climate justice to be able to deal with the complexity of climate change. By challenging our ability to determine what climate justice requires of us, the complex nature of climate change also makes it more difficult for us to determine who does and does not fall within the scope of climate justice. In particular, it challenges our ability to determine whether and why we have duties of climate justice to future people. To see why this is so, we need to unpack what exactly the complexity of climate change consists of. As Jamieson suggests, the problem lies in the convergence of a number of individually tough problems (Jamieson 2014). In light of a fast-changing climate we now face potentially very severe or even catastrophic future outcomes (*severity*)[1] but are still uncertain about whether and when these outcomes will occur and how extensive they will be (*uncertainty*); like its causes, the impacts of climate change are dispersed across space and time, which means people across the globe from now until the very remote future are going to be affected (*dispersion of causes and effects*); at the same time, there are significant differences in agents' contributions and vulnerability to climate change (*asymmetric vulnerabilities*); finally, due to its sheer scale, climate change and the socio-economic changes it will lead to are likely to significantly reshape our societies by affecting who will be born (*scale*).

Many of our most common moral theories struggle to provide an account of the scope of climate justice that can deal with this distinctive set of problems and their convergence. In Chapter 2, I argued that these difficulties are particularly apparent when looking at the consequentialist account of scope, especially with regard to the uncertainty about the future impacts of climate change. Earlier I discussed Jamieson's suggestion that the dispersion of the causes and effects of climate change across time and space makes it, though not impossible, at least very challenging to apportion responsibility for climate harms and consequently to assign the corresponding duties. Similarly, theories of justice that ground duties on cooperation and reciprocity between persons face significant hurdles in the intergenerational setting (see Gardiner 2003, 2009; Page 2006: chapter 5).

Lacking an account of the scope of climate justice that can deal with its complex nature may make us vulnerable to the challenges identified by Gardiner as part of his theoretical storm narrative. That is, not being able to access a robust account of scope may on the one hand tempt us to assume we have no obligations of climate justice to future people. On the other hand, if Gardiner is right to also worry about moral corruption, the lack of clear norms may in turn lead us to overemphasise factors that justify limiting the scope of our obligations in time, such as uncertainty about the future impacts of our actions or the difficulty in identifying and quantifying the harm they cause.

In order to meet the intergenerational challenge, it is therefore crucial that an account of scope be able to provide robust guidance in the face of not just the multiple challenges of climate change but also its overall complexity. In other words, it must meet criterion C2 and be able to respond to climate change as a complex problem of justice. The remainder of this chapter will focus on how we may be able to build an account of scope that meets this requirement. In the next section, I will argue that in order to construct such an account we ought to open up our theories to greater pluralism and diverse methods. I will implement this suggestion in the following section, suggesting that we extend the approach to justice under uncertainty proposed in Chapter 2 into a full account of scope and arguing that the resulting action-centred account would allow us to successfully overcome the climate challenge. I will then summarise the ways in which the proposed approach addresses the key issues identified in this chapter.

The road to a solution

In order to formulate a successful account of scope we have to take seriously the concerns voiced by Jamieson and Gardiner. Because this book focuses on the scope of climate justice, I will only address their arguments as they apply to the question of scope. I will not discuss whether our morality as a whole needs to be developed in relevant areas to meet the climate challenge, as suggested by Gardiner, or whether, as suggested by Jamieson, we need to devise new moral values and concepts to fill in the substance of climate justice.

I will address Gardiner's claim first. There are two reasons why I consider his arguments prior to Jamieson's. The first reason is that responding to Gardiner's interpretation of the problem is less revisionary of our morality. Recall that Gardiner suggests we can solve the struggle of climate justice with complexity by developing our moral theories in the relevant areas, allowing us to access the necessary moral norms. Adapting the values and norms we already have in ways that meet the climate challenge is not only a more straightforward task than devising and justifying a new set of moral values, it also requires less changes to the moral framework we are familiar with. Although it may require us to extend it, developing our moral theories will not require us to abandon the understanding of morality to which we have grown accustomed. It makes sense to test whether this strategy can be successful before embarking on the larger task of revising our entire moral frame of reference.

The second reason is that, if we can show that the values and norms we already subscribe to also apply to future generations in complex situations, the arguments we can make in favour of doing so will be stronger than arguments asking people to adopt a new understanding of morality. If we take the less revisionary route, we only need to call on people who are already committed to these values, and who readily apply them to other areas of life, to extend them to cover climate change and future generations. This places a lighter burden on them than adopting new values. In addition to being the theoretically preferred solution, this may in turn increase people's motivation to act on these principles.[2]

How, then, do we assess whether our values and norms can be developed and extended in the necessary ways, that is, in ways that explain whether and why we owe duties of climate justice to future people? A promising strategy, proposed by Edward Page, is to approach and combine familiar norms through new and diverse methods. As part of his "tentative proposal for greater pluralism," Page argues that theorists like Jamieson have underestimated the potential for using our existing values to moralise climate changing behaviour. He suggests we may in fact be able to do so by "bringing together values that already command widespread allegiance but which are usually viewed as being incompatible." The core of his argument is that, as we try to adapt our philosophy to the demands of climate justice, our theories stand to benefit from greater pluralism and that what we need in order to meet the challenges of climate change are not new values but new methods to use the values we have more creatively (Page 2006: 166–167).

Page's suggestion requires greater examination, for he does not fully explain what the use of new methods and a commitment to greater pluralism may entail. Partly for this reason it is an interesting and promising starting point for my analysis. In what follows, as I attempt to develop a successful account of the scope of climate justice, I will build on his suggestion to approach and combine familiar values by using diverse and new – at least to the field of climate justice – methods. For the sake of my investigation, I understand the appeal to greater pluralism as an encouragement to use one or more methods to combine diverse moral values as well as elements of different approaches to justice, in ways that are better suited to respond to the climate challenge than traditional concepts of justice. In the following section, I will apply this approach to the problem of complexity.

An action-centred solution

Justice, uncertainty, and complexity

In Chapter 2, I argued that an action-centred methodology that takes actions as the subject of moral evaluation and grounds our obligations of justice on the character of our actions – most importantly the presuppositions on which our actions are based – can successfully respond to the problems posed by uncertainty in an action-guiding and normatively accurate manner. This approach allows us, for instance, to clearly establish duties of justice to future people even when we cannot predict the exact impacts of our actions or assign probabilities to their possible effects. It also does two more things that are relevant to our discussion of pluralism as a response to the complexity challenge: on the one hand, it emphasises the existing values of autonomy and respect for agents; on the other hand, it combines these values with an action-centred, practical methodology. To my knowledge, this method – which involves viewing the question of scope as best answered in response to a specific, practical problem, and by assessing the actions and underlying presuppositions that constitute that problem (see O'Neill 1996) – has not been applied to intergenerational climate justice before.

Extending this approach to justice under uncertainty and formalising it into a full account of scope seems promising in response to the complexity challenge and, if successful, will allow us to respond to it without, as Jamieson argues, having to develop new values and moral concepts. What, then, does it take to extend the sketched approach into a full account of scope? In the remainder of this section, I am going to argue for three things. First, I will show that we engage in a number of actions which are based on presuppositions about future people – *intergenerational actions* – including actions on foreseeable risks but also many others. Some of these actions require the contribution of future people in order to be completed successfully, which I will suggest grounds an element of intergenerational cooperation. Second, I will formalise the requirement to recognise as agents those others on which our actions are based by defending a *coherence requirement*. Finally, I will formalise the requirement to respect each agent and their conditions of autonomy by introducing a *normative requirement*. In this chapter I am only going to formulate an initial version of the account with a focus on its ability to overcome the complexity challenge. In Chapter 5, I will then revisit the account to assess how it meets our three criteria for a successful account of scope, clarify three issues left unresolved by the preliminary account, and propose a final amended version.

Collective agents, intentions, and assumptions

Recall that, throughout this book, I focus on *collective agents*. More specifically, I am concerned with those large-scale, organised collectives that are relevant to climate justice. These are all organised collectives engaged in actions relevant to climate justice, that is, in actions that directly affect the extent to which we are able to mitigate, adapt, or otherwise respond to anthropogenic climate change. These actions include both actions and omissions consciously performed and intended by the agents, including attitudes, policies, and practices. I understand this set of collectives to include but not necessarily be limited to nation-states, subnational governments, international organisations, as well as private corporations, research institutions, civil society, and nongovernmental organisations.

As the backdrop to my arguments, I assume what I will call an agential theory of collective responsibility (see Fleming 2017).[3] According to the agential account, states and other collectives that meet a set of specified criteria are agents in all morally relevant respects, over and above the individual agents who constitute the collective. As such, these collectives are capable of forming intentions to engage in specific actions, which are not reducible to the sum of its members' intentions. Some proponents of this type of view are Peter French (1979), Onora O'Neill (1986, 2016), Robert Goodin (1995), Christopher Kutz (2000), and Toni Erskine (2001).[4]

Not all groups, however, are organised collectives in the sense relevant to my arguments. Erskine (2014) summarises five conditions for a group to count as a collective moral agent. It must have

1　a corporate identity, that is, an identity which is more than the identities of its members;
2　a decision-making structure through which it can decide on actions that differ from all or some of its members' positions;
3　mechanisms that can put its decisions into action;
4　an identity which persists over time; and
5　an internal conception of itself as a unit, that is, the group cannot be only externally defined.[5]

Groups that meet these conditions are, for example, most functioning democratic states, private corporations, international and intergovernmental organisations, and many civil society groups. According to the agential view, these groups are moral agents capable of exercising agency, assuming moral responsibilities, and forming a "collective will" (Valentini 2011: 133) as well as corporate intentions (French 1984).

Because of their corporate organisational structure and decision-making mechanisms, these agents, like individuals, "bring cognitive and decision-making capacities and capabilities to bear on choices that initiate action and affect what happens" (O'Neill 2016: 167). They may come to have reasons for action, or intentions, which differ from the reasons of its individual members (French 1979: 214; Kutz 2000: 97–103).

Because my focus lies on collective agents, I am only going to consider the intentions that are attributable to collective agents in such an institutionally embedded context and not the underlying individual intentions. Duties of climate justice that follow from my account of scope will consequently be attributable to these collective agents. A full account of climate justice will have to answer a separate set of questions on whether, and if so how, these duties ought to be distributed among individual members, whether individuals incur additional duties of climate justice, and if so what the scope of these duties is. Though important, these questions are outside the scope of this book, and I will set them aside for the time being.

Discerning the presuppositions behind organised collective actions

As a second and final clarification before I lay out my arguments, let me explain how we may be able to discern the presuppositions behind collective actions. Recall that these presuppositions are not part of the agent's consciousness but rather a feature of activity and part of the character of our actions. The presuppositions of activity I am interested in are those about the capabilities and vulnerabilities of others who may be affected by the action. In the same way as they can be attributed to individual actions, such presuppositions are also involved in the other-affecting action of collective agents. Consider, for example, the practice of public debt. A state taking on public debt is based on the presupposition that there are going to be future generations of citizens to repay it. The fact that states can and do take on long-term debt is partly explained by this presupposition about future people. In turn, this presupposition shapes the action, for example, by allowing for a longer time frame for debt restitution. Without the presupposition that there are going to be future citizens, time frames for repayment would plausibly be much more limited in time. Similarly, actions of corporations are often based on presuppositions about their employees or distant agents. The act of guaranteeing employee benefits and vacation days, for instance, presupposes employees as agents who stand to be affected by the corporation's actions, who may leave if they perceive themselves to be treated unjustly and denied periods of break, and who will perform poorly if exhausted.

Actions may also be based on presuppositions about agents that can be specified but not identified (see O'Neill 1996: 113–121). Consider a company's act of paying for externally provided products or services. This presupposes external agents who have the capabilities to provide these products or services in exchange for money. A company setting up a pension scheme for its employees, too, presupposes a number of things about unidentified others. Investing employee contributions on the stock market presupposes the existence of other agents who also sell and buy stock, and depending on the scheme's structure, it may be based on the presupposition that future employees will contribute to funding the scheme by participating in it themselves. As these examples suggest, our actions are often based on presuppositions about others whose identity we neither know nor need to know.

For the moment, what matters is the empirical claim that other-affecting action by collective agents involves presuppositions about others. How these presuppositions affect the scope of agents' duties will be explained by the coherence and normative requirements I will argue for later. These requirements state respectively that if our actions are based on presuppositions about other agents, we also ought to recognise these others in our principle of justice and that doing so requires us to extend the basic requirements of justice to them, that is, respect their entitlements to the conditions of autonomy. If and when collective agents relevant to climate justice engage in actions that are based on presuppositions about future people, they therefore acquire relational, action-based duties of justice, including of climate justice, to those future others. The presuppositions behind our actions thus form the basis of the account of the scope of climate justice I am going to propose.

One could ask whether we run the risk of incorrectly identifying the presuppositions on which an action is based, in ways that would affect the resulting moral obligations. Melissa Barry, commenting on Onora O'Neill's account of justice and virtue, writes that often, the presuppositions behind a certain activity may not be obvious. As an example, she cites an agent's attempt at cheating another with a fraudulent investment. She argues there are two possible ways of interpreting the character of the action: on the one hand, it could be based on the presupposition that the victim is a rational agent able to follow the reasoning the agent offers in support of the (fraudulent) scheme. On the other hand, the cheating attempt could also be based on the presupposition that the other is rational yet gullible enough to fall for the fraud. Both presuppositions can explain why the agent acts as she does. Barry therefore argues that if the second, more fine-grained proposal is in fact the more plausible way to describe the relevant presuppositions, then the argument fails to show that the agent ought to treat the victim with respect. This is because the agent's actions do not rely on presuppositions about the victim as a fully but only as a partially rational agent (Barry 2013: 26–27).

Barry is right to argue that, depending on how fine-grained the presuppositions are that we identify, it may in certain cases be unclear what amount of rationality or other capabilities our actions presuppose. However, on the account I am going to argue for, it is irrelevant which degree of a capability our actions presuppose. The relevant work in grounding our duty to respect the agents whom our actions presuppose is done by the coherence and normative requirements, both of which are based only on possessing a certain capability, for example, for human agency. This is understood as a binary property rather than a matter of degree. That is, we do not need to presuppose a certain degree of agency in a person for her to be owed respect as an agent. I will argue in Chapter 5 that we may owe more or less specific duties to others depending on which other capabilities, beyond a basic capability for agency, our actions presuppose in them. For all relevant capabilities, however, as for agency, it suffices that our actions presuppose the agent to have that capability. None of the duties I will argue for rely on overly fine-grained presuppositions.[6]

Intergenerational actions

The presuppositions of activity are relevant to the intergenerational scope of climate justice because they show that a number of present actions involve presuppositions about the agency of future people. In other words, they show that many actions by collective agents relevant to climate justice are *intergenerational actions*.

> INTERGENERATIONAL ACTIONS: *Actions that are based on presuppositions about the capabilities and vulnerabilities of future people.*

Actions can be based on variously strong presuppositions about future people. For the moment, we may think of intergenerational actions in two categories: strong intergenerational actions, which rely on the contribution of future people to be completed successfully and are therefore based on particularly strong presuppositions about future persons – these actions, as I will argue, ground an element of intergenerational cooperation; and weaker intergenerational actions, which do not require future people to contribute in order to be completed successfully but which nevertheless rely on presuppositions about future persons. While I will argue in Chapter 5 that this strict dichotomy fails to include actions that lie between strong and weak intergenerational actions and that intergenerationality is instead better conceived as a matter of degree, this is a useful distinction to make while I lay out the general argument. I will discuss strong and weak intergenerational actions in turn.

Strong intergenerational actions cannot be completed successfully without the contribution of future people. In this sense, they enable non-overlapping generations to cooperate with each other through their respective engagement in the same activity. There are several historical examples of this category of actions. Some obvious ones are the construction of great landmarks or religious sites, which often took more than one or even two generations' lifetimes to be completed. The construction of Notre-Dame de Paris, for example, began in 1163 and lasted nearly two centuries (Association Maurice de Sully 2019). It is clear to see why this is an intergenerational action: starting what one knows to be such a lengthy construction project clearly relies on the presupposition that there are going to be future people capable of bringing it to completion. Importantly, the presupposition need not be that future people are going to successfully complete the project; rather, the presuppositions of strong intergenerational actions are that future people are going to have the capabilities necessary to bring the project to completion and the willingness to use them for this common endeavour.

As a strong intergenerational action, the construction of Notre-Dame thus relies on a very specific presupposition about the future: it not only presupposes that there will be future people but that there will be future people able and willing to continue that specific project. It relies on such a specific presupposition because the action is such that it cannot be completed unless such people exist and contribute. In other words, without the presupposition that there are going to be future people capable and willing to continue construction, the act of starting such a long-term construction project seems to be conceptually incoherent. In turn, the presupposition that there are going to be such future persons explains why present people are willing to begin an activity they know requires more than one generation to be completed. In the case of intergenerational construction projects like Notre-Dame, in which each generation's contribution is clearly visible, the intergenerational cooperative element of more than one generation contributing to the same project is particularly tangible.

Strong intergenerational actions also abound in contemporary societies, to the extent that many see our societies as intrinsically intergenerational (Andina 2018a, 2018b; Skinner 2009: 364; Thompson 2002). According to Tiziana

Andina, our societies rely heavily on the existence of transgenerational connections of trust, expectations, and cooperation, embodied in such intergenerational actions (Andina 2018a, 2018b). Similarly, Janna Thompson puts forward a view of societies and nations as intergenerational communities that are dependent on their members accepting transgenerational obligations. Members of these societies also make demands on their successors, including incurring commitments they see as binding posterity and which they believe their successors ought to honour (Thompson 2002: xvii).

My earlier example of public debt, for example, fits the definition of a strong intergenerational action. The practice of public debt involves lending money to one generation which is then to be repaid by future citizens of that country. This practice is based on the presupposition that there are going to be future citizens, that these citizens and their governments are at least likely to have the capabilities to repay the loan or otherwise integrate it in their financial planning, and that they will be willing to do so. Like the construction of Notre-Dame, taking on public debt is strongly intergenerational for it involves one generation engaging in an action – in this case, participating in a practice – that can only be successfully completed with the contribution of future generations, and which is therefore based on the presupposition that future people will be able and willing to cooperate. Again, this presupposition is essential to the activity: given that its implementation requires the contribution of more than one generation, the practice is only conceptually coherent given the presupposition that there are going to be future agents able to contribute so that the activity may be completed successfully.[7]

Actions related to climate justice are often intergenerational in this sense, too. Consider, for example, national policies to pursue a specific mitigation path or enact a climate action plan. A number of these policies were devised by states as a means to implement the mitigation pledges published in their Nationally Determined Contributions (NDCs) under the United Nations Framework Convention on Climate Change (UNFCCC).[8] These policies set out long-term mitigation, adaptation, and investment plans in line with the time horizons of the Paris Agreement. Germany's Climate Action Plan 2050, published shortly after the Paris Agreement, is one example of this. The plan sets out emission reduction targets for 2050 and 2030, as well as sectoral targets, implementation strategies, and individual measures up to 2030 in sectors like buildings and transport, industry and business, agriculture and forestry, and energy supply (Bundesministerium für Umwelt, Naturschutz und nukleare Sicherheit 2016). A second climate package, setting out emission reduction targets for 2030 and including a Climate Action Law which enshrines these into law, was passed by the government in December 2019 (Appunn and Wettengel 2019; Deutscher Bundestag 2019; Wehrmann 2019). Both sets of targets and measures span further than the current government and will have to be implemented in cooperation with not just a new administration but also future generations of policymakers representing future generations of citizens. Implementing the kind of long-term targets, roadmaps, and measures set out in the plan is an action that cannot be completed successfully unless future generations cooperate. In turn, as in my previous examples, these medium- and long-term policies presuppose that future people will be able and willing to contribute and are at least likely to do so.

The same applies to many more long-term policies enacted by nations, communities, or other collective agents. A number of these actions are being taken in the context of climate change, and although a close analysis of all is beyond the scope of this book, it is worth mentioning a few relevant examples. These are, for instance, the emission reduction goals and climate action plans that are being adopted by cities and regions all over the world. Cape Town, Barcelona, and New York, to name a few, have all committed to be carbon neutral by 2050 (C40 2017, 2018, 2020; CDP 2021), and the state of New York, too, adopted its own climate change plan. The plan commits the state to reducing its greenhouse gas emissions by 85% below 1990 levels by 2050 and to producing 40% of its electricity from renewable sources by 2040 (Bekiempis 2019; Storrow 2019). It is important to emphasise that my argument does not depend on whether these policies will in fact be implemented by succeeding generations, in other words on whether the initiated intergenerational action will be successful. What matters is that the agent's activity at the time of action is based on the presupposition that there will be future people who will be able, willing, and at least likely to continue it.

Actions related to climate justice are particularly interesting examples of intergenerational actions. On the one hand, they are relevant examples of general actions, including public actions, based on presuppositions about future agents. On the other hand, they exemplify the paradox of mainstream responses to climate change. Through their various responses and the underlying intergenerational presuppositions, agents relevant to climate justice are connected to future agents. At the same time, they seem unaware of these intergenerational connections and their implications for climate justice. These connections not only commit agents to include future people within the scope of their obligations but are in most cases also likely, on a substantive level, to require greater climate action for future generations. There is hope that raising awareness about this incoherence may motivate some degree of change, even if small, towards greater protection of the environment for future people.

These activities are only some examples of strong intergenerational actions. They exist within a deeply intergenerational understanding of societies. That agents engage in strong, cooperative actions at all shows that we understand societies as intrinsically intergenerational constructs in which there is a real possibility for strong intergenerational actions to succeed. This is important for it suggests that there are a great deal of actions, beyond the examples mentioned here, which are in fact intergenerational and hence give rise to intergenerational duties. Most importantly, actions shaped by presuppositions about future people are not exclusive to the sphere of climate justice. Many activities that play a vital role in the functioning of our societies – many investment, planning, or policy decisions – are based on presuppositions about the future, and a number of these rely on future people to be completed successfully. The practice of public debt is only one example. Others are long-term public or private projects such as long-running space programmes, large infrastructure projects, or the establishment of and investment in public or private pension funds. The abundance of intergenerational actions suggests not only that they play a significant role in contemporary societies but also that most agents relevant to climate justice are likely to engage in one and often multiple intergenerational actions.

I have now explained what strong intergenerational actions are. Weak intergenerational actions, on the other hand, are actions that are based on presuppositions about future people but do not rely on their contribution for success. Let me give you some examples. One core category of weak intergenerational actions, discussed in Chapter 2, is actions that impose foreseeable risks on future generations. I argued that these actions, because they impose foreseeable risks on future people, necessarily involve weighing the preferences of the risk-imposing agent against those of the risk-exposed. They therefore involve presuppositions about the capabilities and vulnerabilities of future people and how they stand to be affected by the risk. Examples of these activities are actions taken in response to the risk of the large-scale melting of the WAIS or in response to the risk of extreme sea-level rise due to climate change. Other examples are actions related to the production, handling, and storage of nuclear waste and the long-term risks it imposes, as well as activities related to research into potentially destructive weapons technology. Because I explored this typology of actions at length in Chapter 2, I will set them aside for now.

The second category of weak intergenerational actions is not directly linked to a risk-imposition yet still involves presuppositions about future persons. Having looked at national climate action plans discussed earlier, I now want to explore an international legal document on climate change as an interesting example of such a weak action. More specifically, I will argue that the Paris Agreement signed in 2015 relies on presuppositions about future people and that by authoring and signing the Agreement, parties to it engaged in an intergenerational action.

The Paris Agreement sets a binding target, agreed consensually by all parties to the Agreement, to "hold . . . the increase in the global average temperature to well below 2°C above pre-industrial levels and pursu[e] efforts to limit the temperature increase to 1.5°C above pre-industrial levels" (United Nations Climate Change 2015). In order to achieve either temperature goal, we have to pursue mitigation pathways that allow us to gradually limit the amount of greenhouse gases present in the atmosphere. Possible trajectories are regularly calculated by the IPCC and published in their periodical assessment reports. In what is at the time of writing its latest report on climate change mitigation, the IPCC states that in order for it to be likely that temperature increase in 2100 be less than 2°C, atmospheric concentration levels ought to reach approximately 450 ppm CO_2eq in 2100, with the possibility of a temporary overshoot before 2100.[9] The IPCC also lays out the mitigation efforts required to reach a given concentration level. Scenarios leading to levels of about 450 ppm CO_2eq require substantial emission cuts by 2050, followed by further cuts to bring emission levels close to or below zero in 2100 (IPCC 2014: 10). Typically, scenarios that allow us to reach these concentrations also involve a temporary overshoot in concentration levels before 2100. According to the IPCC predictions, correcting these overshoots will require the deployment of carbon dioxide removal (CDR) technologies. As the IPCC also notes, CDR technologies have their own risks and are currently unavailable at the necessary scale (IPCC 2014: 12).

This information was available at the time the Paris Agreement was drawn up and is very likely to have informed the foregoing negotiations. Especially given that parties to the Agreement knew of the intergenerational efforts that would be required to achieve it, setting a future-oriented mitigation goal as is entailed in the Paris Agreement clearly presupposes the existence of future agents. More specifically, it presupposes future people who are, on the one hand, still vulnerable to the effects of climate change and thus at least likely to wish to continue efforts to mitigate its effects;[10] and who, on the other hand, are at least likely to have the capabilities needed to take the required measures. In other words, the Agreement's future-oriented mitigation target and related policies are only conceptually coherent given the presupposition that such people are going to exist. As they authored, signed, and ratified the Agreement – the text of which was agreed to consensually by all parties – participating states thus engaged in an intergenerational action.[11]

The Agreement contains other intergenerational elements which further expose its authoring, signing, and ratifying as intergenerational actions. It includes, for example, a periodical review mechanism for NDCs. This mechanism is critical if the Agreement is to be at all successful, especially since current climate policies leave a substantial emission gap. What this means is that, taken together, all emission reductions currently in place only keep us on track for a warming of 2.7°C. Factoring in current pledges and mitigation target would bring us to somewhat less worrying, but still insufficient, 2.1°C (Climate Analytics and New Climate Institute 2021). Research also shows that atmospheric concentration levels of greenhouse gases had already reached 449 ppm CO_2eq in 2016, with current emission rates suggesting they are unlikely to peak within the next decade (European Environmental Agency 2019; United Nations Environment Programme 2019). In order to progress from these NDCs towards the mitigation objectives of the Paris Agreement, states built into the accord a commitment to review and ratchet up their NDCs every 5 years, setting no upper time limit (Taibi and Konrad 2018). The review mechanism is a way to motivate and track progress towards the Agreement's temperature goal and highlights an additional way in which the Agreement presupposes the existence of future agents. Just like setting a future-oriented temperature goal, including a periodical and hence future-oriented review mechanism in the Agreement presupposes at the very least a real possibility that, as the review mechanism progresses, future governments representing future citizens will be able to review and ratchet up their NDCs. That is, including such a mechanism that is unlimited in time would be conceptually incoherent without the presupposition that there are going to be the appropriate future agents.

Later I will consider three objections to the view of presuppositions behind collective actions I have just laid out. Before I do that, in the following pages I am going to introduce the coherence and normative requirements that complete my account of scope.

Intergenerational obligations

Intergenerational actions are the empirical basis for the account of scope I propose. These actions impose intergenerational obligations of justice on their
agents – including obligations of climate justice, where the agents are relevant to
climate justice – through a *coherence requirement* and a *normative requirement*.
The coherence requirement holds that because our actions rely on presuppositions about future people as agents, we have an obligation to acknowledge their
standing as agents when thinking about justice. The normative requirement holds
that insofar as we are committed to basic norms of justice requiring equal respect
for agents and their conditions of autonomy, we must extend these entitlements
to future agents presupposed by our actions. In other words, we ought to include
these future agents within the scope of our obligations of (climate) justice. I will
address each requirement in turn.

The *coherence requirement* is the claim that, whenever agents' own actions
are based on presuppositions about the capabilities and vulnerabilities of others,
agents have an obligation to recognise these others' agency when thinking about
the obligations of justice that ought to govern their actions. My discussion of
intergenerational actions showed that many collective actions in contemporary
societies rely on presuppositions about the capabilities and vulnerabilities of
future people. I suggested that this is the case for actions related to climate justice
as for a wide range of actions in other areas of social, political, and economic life.
Regardless of how strongly intergenerational these actions are, they have in common the presuppositions that there will be future people with certain capabilities
and vulnerabilities who can be affected by these actions and who will be able to
react to them. These actions are only conceptually coherent given the underlying
presuppositions about future others as agents with capabilities and vulnerabilities.
This implies that agents could not engage in these actions in their current form
were it not for the underlying presuppositions about future others as agents.

That agents' actions involve such presuppositions about others' agency imposes
an obligation on them to respect these others as agents. On an understanding of
justice in which respect for agency, autonomy, and the equal moral status of persons are key, agents would engage in a practical contradiction if they were to
act in ways that presuppose others yet fail to recognise and respect these others
as agents. In other words, if agents make use of, and thereby expect, respect for
their own agency yet deny others whom their own actions recognise as agents
with certain capabilities and vulnerabilities the same respect for their agency, they
are guilty of making an unwarranted exception of themselves. The central claim
of the coherence requirement is therefore that, once they have committed to presuppositions about others as agents through their very own actions, agents may
not deny these presuppositions of agency when devising principles of justice. If
agents engage in intergenerational actions, their principles of justice and climate
justice ought to recognise the standing of future people as agents, too.[12]

The second requirement, and third and final component of my account of scope,
is a *normative requirement*. In Chapter 2, I argued that each agent is entitled to
the basic conditions of autonomy and that others have a corresponding duty to

respect these entitlements. The normative requirement draws on these existing values to spell out what it means to recognise and respect others' agency, as the coherence requirement requires us to do. It holds that recognising and respecting the agency of others entails an obligation for agents to respect these others' entitlements to the conditions of autonomy. Agents who engage in intergenerational actions therefore acquire an obligation to act in ways that respect the autonomy conditions of the future agents presupposed by their actions.

Through our intergenerational actions, the coherence and normative requirements thus extend our familiar values of respect and autonomy to future agents. The agential theory of collective responsibility laid out earlier implies that both requirements apply to organised collectives as to individual agents. Hence, whenever the actions of organised collective agents are based on presuppositions about others' agency, these collective agents acquire an obligation to respect those others and their autonomy. Because intergenerational actions are so widespread, we can assume that all or very nearly all collective agents relevant to climate justice are engaged in intergenerational actions and that, accordingly, the scope of their obligations of (climate) justice includes future agents.

It is important to note that the value of respect for agency, which has been a guiding thread throughout this book, and which is embodied by the coherence and normative requirements, is in line with a specifically Kantian interpretation of morality.[13] As such, it represents only one of many possible views of moral philosophy. As one of the established currents within moral philosophy it is likely to have at least as many opponents as supporters, but in the limited space of this book I will not attempt to defend the view itself against their objections. Nevertheless, while I take for granted the basic principles of this specific view, I do hope that some of the arguments I build on it may also speak to some of its opponents.

Because the principle of respect for agency I rely on is part of a broader view of moral philosophy, accepting it as part of one's overall framework of morality will have implications beyond those related to the account of the scope of climate justice I defend. A view of morality based on respect for agency may also, for example, impose negative or positive duties on agents even when their actions do not rely on presuppositions about others. The practical account of scope I defend in this book only covers one part of our moral obligations and only in response to the specific problem of climate change. Even within the domain of climate justice, it determines the scope of only one set of duties that are intergenerational in reach. It shows that, and why, the scope of at least some duties of climate justice is intergenerational but does not rule out that we may have additional, complementary duties of climate justice that are not related to intergenerational actions and which may or may not be of intergenerational scope. In other words, the account I propose is itself limited in scope, as is the purpose of this book. And while the broader principles on which it relies will have implications beyond it that are likely to be opposed by those who subscribe to a different view of moral philosophy, discussing these implications falls beyond the scope of this investigation.

These limitations aside, the account I have sketched successfully answers the question of whether, and if so why, we have duties of climate justice to future people and is able to do so in spite of the complexity of climate change. It does so by grounding our intergenerational duties in the actions we are already engaging in and therefore cannot deny and which are accessible to us as a source of information even in a situation as complex as climate change. Because of these traits, arguments for extending the scope of justice to future agents based on this account seem particularly strong. These arguments provide us with additional reasons, beyond those given by existing accounts of intergenerational justice, why we ought to include future people within the scope of climate justice.

Scrutinising the presuppositions behind collective actions

I will now consider three possible objections to how I argued we are able to discern the presuppositions behind the actions of collective agents. The first objection is as follows: it is not necessarily true that, as an individual participating in a collective action, I intend that its collective goal be successful. This is implied by my definition of collective action: the agential theory I proposed holds that collective intentions can differ from the intentions of some or all of its members where the collective agent is capable of a collective will formation. This affects my argument, as the act description used to discern the presuppositions underlying collective action may no longer be appropriate once the differing intentions of the participants are taken into account.

A response to this objection, too, can draw on the definition of collective agency outlined earlier. The agential theory states that collective intentions can exist over and above individual intentions. Accordingly, my argument is based on the collective action resulting from that collective intention. It grounds the resulting duties on that action's underlying presuppositions and attributes them to the collective agents responsible for it. That the intentions of individual members of these collectives – and hence the appropriate act description for their individual actions, where these can be differentiated from the collective act – can differ from the collective intention is correct. However, whenever we deal with organised collectives that meet the criteria set out earlier, it is of no importance whether individual citizens of a state, or members of a corporation or organisation, intend for the collective goal to succeed. It is the collective agent, to whom the collective intention and the resulting action can be attributed, who acquires the resulting obligations. In the cases discussed, we can clearly attribute to states the acts (and consequently the intention) of signing the Paris Agreement or enacting certain climate policies, and we can do so regardless of their underlying motives or their citizens' individual intentions.

The second objection is similar but is now raised at the UNFCCC level: does it matter which individual motives states were pursuing in authoring and signing the Paris Agreement, if, for example, their participation was a purely political move with no underlying commitment to the Agreement, its policies, and goals? To be clear, this objection does not question member states' intention to author and sign the agreement. The intention is clear from the fact that states performed these actions: they clearly intended to author and sign that very document. What it does question are states' possibly insincere motives for doing so and whether these affect the underlying presuppositions and hence the duties that can be derived from their actions.

The short answer is no: what I have argued so far suggests that agents' motives for an action do not affect the duties they acquire by engaging in it. Recall that the presuppositions on which I based my argument are a feature of the relevant action rather than part of the agent's consciousness or motives and that they may at times not be acknowledged or even actively denied by the agent. Drawing on these pre-conditions of action, I argued that agents whose actions presuppose others have an obligation to recognise these others as agents and respect their entitlements to the conditions of autonomy. The resulting duties are derived directly from someone's actions and independently of their motivations for acting. In the case of the Paris Agreement, the actions in question are the negotiating, signing, and ratifying of the Agreement, which I have argued presuppose the agency of future people in various ways. Now, imagine state A intended to take these actions not as a sincere commitment but as a political stunt to appease internal and external opposition to its climate denialist policies. The action A needs to take to appease its opposition is to sign a future-oriented accord. I have argued that such a future-oriented accord would not be conceptually coherent without the underlying presupposition that there are going to be future people. In turn, this means that regardless of why A signs the Agreement, their doing so is based on presuppositions about future people. This is so whether A acknowledges or denies these presuppositions. Engaging in any action that relies on presuppositions about future people, for whatever purpose, thus triggers the coherence and normative requirements. In other words, taking the same action in pursuit of a covert aim, even if in opposition to the action's official purpose, does not relieve agents of the duties the action gives rise to.

The third and final objection questions an underlying premise of my arguments. It asks whether, in situations of deep uncertainty, collective agents are at all able to engage in the type of long-term or future-oriented action that presupposes future persons. This objection combines two separate worries: first, a general scepticism about the ability of collective agents to engage in long-term and future-oriented action; second, the question of whether their ability to do so is especially constrained in situations of deep uncertainty. I will consider each point in turn.

On the agential theory of collective responsibility, organised collective agents that meet the relevant criteria are seen as entities that are, among others, capable of forming a collective will. One of the necessary criteria is possessing an identity as a collective agent which persists over time, that is, being a temporally extended agent. A straightforward example of a collective agent that meets these criteria, and which is therefore seen as able to form a collective will, is the modern democratic state. If we accept the agential view of collective responsibility and agree that collectives can be agents over and above its individual members and are able to form a collective will, there is no reason why this capacity should not include the ability to engage long-term and future-oriented action. On the contrary, it seems even more plausible that temporally extended collective agents such as states, as opposed to individual agents with much shorter lifespans, be able to and indeed would engage in long-term and future-oriented activity.

In its second part, the objection questions whether this can be true even in contexts of deep uncertainty such as climate change. This worry seems to rest on a misunderstanding about future-oriented, intergenerational actions. It seems to wrongly assume that, in order for an action to be based on presuppositions about the future, its agents need to be able to predict the future and the future effects of their actions with some degree of certainty. This is incorrect. I argued earlier that engaging in cooperative intergenerational actions, for instance, confers intergenerational obligations on agents regardless of whether the action is completed successfully. Agents acquire obligations whenever their present actions are shaped by certain presuppositions bout future people. Even in contexts of uncertainty, it is reasonable for these kinds of presuppositions – that there are going to be future agents with certain capabilities and vulnerabilities – to form the backbone of our temporally extended, intergenerational societies and in turn frame many of our actions. Such presuppositions are necessary in order for agents to be able to act at all in the many situations of uncertainty which permeate our lives.

The prevalence of uncertainty in the context of climate justice means, on the one hand, that while agents have to act on uncertain predictions they nevertheless rely on presuppositions about future people when they engage in actions that impose uncertain yet foreseeable future risks. This is what I argued in Chapter 2. On the other hand, uncertainty about future climate change also does not preclude agents from engaging in additional, future-oriented, and climate justice–related actions that are not centred on risk. Consider the German climate action plans discussed earlier. They can and have been heavily criticised for not doing enough to slow climate change[14] and may in hindsight turn out to have been inadequate policies for the threat we are facing. Yet despite how accurate or inaccurate their underlying predictions may be, or which predictions they may have turned a blind eye to, these are nevertheless intergenerational actions based on the presuppositions that future people are going to be affected by and able to respond to them in various ways. Whether these predictions and presuppositions turn out to have been correct is a separate question to whether present actions rely on them.

Those still unconvinced by my arguments on collective, future-oriented action under deep uncertainty can look outside the domain of climate justice to find a variety of alternative intergenerational actions which also ground collective, intergenerational obligations. As argued earlier, collective agents commonly engage in actions based on presuppositions about future people in many areas of contemporary social, economic, and political life where uncertainty is less pronounced than in climate policy. These actions include strong as well as weak intergenerational actions. They range from long-term loans to development plans and projects, long-term investments, or economic recovery plans. While I do not have the space to consider these examples in greater detail, in combination with the actions discussed earlier, they emphasise the important role intergenerational actions play in contemporary societies. As most agents relevant to climate justice will engage in at least one, and likely more than one such intergenerational action, readers unconvinced by the possibility of intergenerational collective action under deep uncertainty can rely on actions outside the domain of climate justice as grounds for intergenerational obligations. If one accepts the general structure of my arguments, there are a number of possible actions showing that collective agents relevant to climate justice rely on presuppositions about future persons and that they ought therefore to include these future others within the scope of their obligations of climate justice.

Overcoming the complexity challenge

This chapter sought to propose an approach to fixing the scope of climate justice which, despite the complexity of climate change, can provide normatively accurate and useful practical guidance to agents. Recall that the complexity of climate change consists in the convergence of many individually tough issues – the extent and severity of possible outcomes, uncertainty, the temporal and spatial dispersion of causes and effects, asymmetric vulnerabilities, and its scale – which together make climate change a particularly complex problem of justice. In this section, I will take stock of whether and how the proposed approach can offer a successful response to the complexity challenge.

I argued that the problem of complexity can be overcome by adopting an action-centred account of scope. Despite the complexity of climate change as a moral problem, focusing on the moral evaluation of our actions and their presuppositions shows that we do have obligations of climate justice to future people and explains why this is so. Because this approach takes our actions as the locus of moral concern, we only need to assess our actions and presuppositions in order to determine the scope of our obligations; that is, we only require information that is accessible to agents even in complex, uncertain situations like climate change. Such an action-centred approach is therefore less vulnerable to the challenges of climate change than rival approaches.

Contrary to what Jamieson argues, this approach can overcome the problem of complexity without needing to develop new norms and values. Instead, as suggested by Page, it uses a methodology which is new to the field of climate justice as a tool for applying our existing moral values to intergenerational climate change. This allows us to access the necessary norms of intergenerational climate justice by responding only to Gardiner's more minimal, less revisionary interpretation of the complexity challenge. Recall that, on Jamieson's interpretation of the challenge, a successful response to the complexity of climate change would require new values and concepts that are better suited to deal with this complex moral problem. I suggested that while Jamieson is right in pointing to climate change as a particularly complex moral problem, we should only seek to develop altogether new values as an option of last resort. Instead, we ought first to test whether we can develop our moral theories and repurpose our existing values in a helpful way.

This less invasive strategy is both Gardiner's proposed strategy and what my approach seeks to do. Contrary to Jamieson, Gardiner believes the complexity problem is caused not by our lack of concepts but by our failure to access norms which would allow us to respond to climate change successfully. The reason we lack access to these norms is that our moral theories are worryingly underdeveloped in areas relevant to climate justice, including intergenerational ethics. In this chapter I have attempted to develop our theoretical understanding of climate justice, at least in relation to its scope, so as to allow us to access important norms of intergenerational climate justice. The action-centred methodology I applied demonstrates that many actions by agents relevant to climate justice are intergenerational and are based on presuppositions about the agency of future people. It also shows that, by engaging in these actions, agents rely on presuppositions about future people and should therefore recognise these future others as agents for the sake of climate justice too. I argued that this, in turn, requires them to respect future people and their entitlements to the fundamental conditions of autonomy. Because we are already committed to these norms of justice and respect for agency, this account is comparatively strong – it only calls for present generations to extend the circle of their existing moral obligations, rather than to adopt a new set of commitments.

This chapter has also highlighted that many actions relevant to climate justice are in fact intergenerational. This is particularly interesting as it suggests that climate change is an intrinsically intergenerational phenomenon. While it is old news that climate change is characterised by a temporal element and its considerable extension over time, this account of scope shows that its intergenerationality goes deeper than that. Assessing our actions shows that climate change is caused by, among others, intergenerationally acting agents engaged in emission-intensive intergenerational actions, ranging from infrastructure and transport projects to investments in the fossils fuel sector. At the same time, most of the local, national, and international responses to climate change we are currently seeing are intergenerational in nature. All these actions rely on intergenerational connections based on presuppositions, expectations, trust, and sometimes cooperation with future people. In both its causation and possible pathways to a solution, climate change is deeply intergenerational. It is noteworthy that the very actions we engage in that are contributing to climate change as well as the actions we take in response to it are intergenerational in nature and form part of a deeply intergenerational problem. This realisation may have some potential to shift public attitudes towards climate change by underlining that and in what ways it differs from many problems of justice we are used to deal with. Awareness of the intergenerational nature of climate change can not only improve our understanding of it as a problem of justice but may also pave the way for new approaches to address it.

Conclusion

In this chapter, I argued that the complexity of climate change poses a problem for accounts of the scope of climate justice. In order to overcome this challenge, an account of scope must be able to explain whether, and if so why, we have duties of climate justice to future people despite the complexity of climate change as a moral problem. This is the second criterion for a successful account of scope.

I argued that we can overcome the problem of complexity by adopting an action-centred methodology, which allows us to extend the values of respect and autonomy to apply to intergenerational climate justice. This approach requires us, first, to assess our actions and the presuppositions on which they are based. Doing so shows that collective agents relevant to climate justice commonly engage in intergenerational actions. I argued that actions are intergenerational whenever they rely on presuppositions about the capabilities and vulnerabilities of future people. Many actions by agents relevant to climate justice are intergenerational, both within and outside the domain of climate policy. I showed that some intergenerational actions cannot be completed without the contribution of future people and that these strongly intergenerational actions also ground a form of intergenerational cooperation.

As the second part of the approach, I argued that whenever agents engage in intergenerational actions, they ought to recognise those future others presupposed by their actions as agents for the sake of climate justice too. This, in turn, requires them to extend to future people our existing norms of justice and respect for agency. In other words, agents who engage in intergenerational actions must include future agents within the scope of their obligations of climate justice.

In the following chapter I will introduce the third and final criterion for a successful account of the scope of climate justice. This criterion will address the question of how future people are to be included in the scope of our obligations and holds that a successful account must include future people for their own sake and as ends in themselves.

Notes

1 For some examples, see Hartzell-Nichols 2017; Mann 2015; Oppenheimer et al. 2014; Shankman 2019.
2 For a discussion of the difference between revising our moral principles, extending them to embrace new commitments, and showing that a matter falls under a principle we are already committed to, see Jamieson 2014: 169–170. As outlined earlier and contrary to what I suggest, Jamieson argues that our principles cannot be extended to cover climate change but that we must develop new moral understandings to do so.
3 I do not have space to defend this account here. For more on collective intentionality, see Jankovic and Ludwig 2017.
4 There are important variations between the details of these theorists' views. They have in common the belief that collective intentions are not reducible to individual intentions, which is all I need to extrapolate for the purpose of my argument.
5 Similar, but more minimal, conditions are identified by French 1984, who define collective moral agents as collectives with corporate identities independent of membership and corporate internal decision structures. Kutz argues, along similar lines, that collective intentions are attributable to groups if group membership is intentional and there is a collective decision rule in place, as well as sufficient overlap between the participatory intentions of its members (Kutz 2000: 108). Similar conditions are also proposed by O'Neill, who identifies nation-states as an example of collective agents insofar as they "have structures which provide methods of *integrating* a large range of information in *rational* ways and *enacting* national policies, while maintaining some national *independence* from other agents . . . and forces" (O'Neill 1986: 62).

6 This approach, too, leaves room for disagreement about the capabilities that are pre-supposed by our actions yet considerably less than if the account were to rely on a more fine-grained assessment of the degree of our presuppositions. In most, if not all, cases, the presuppositions behind an action will be obvious and easily backed up by argumentation. In the rare cases in which they are not, disagreements may, for instance, be solved by a process of public deliberation.

7 For more on this point and intergenerational actions in general, see Andina 2018a, 2018b.

8 The Paris Agreement requires Parties to the Agreement to prepare and communicate their Nationally Determined Contributions. NDCs lay out each member state's com-mitments to mitigation and adaptation efforts to achieve the Agreement's goal of limit-ing temperature increase to 2°C above pre-industrial levels (United Nations Climate Change 2015).

9 Concentration levels are expressed in parts per million of CO_2-equivalents. The IPCC defines CO_2-equivalent concentration of greenhouse gases as "the concentration of carbon dioxide (CO_2) that would cause the same radiative forcing as a given mixture of CO_2 and other forcing components" (IPCC 2014: 1257).

10 At least some minimal efforts. What we are currently seeing is that even our extreme vulnerability to climate change has not motivated states to go much further than paying lip service to their Agreement, thereby widely missing the mark in terms of what an effective and equitable climate strategy would require.

11 At the time of writing, 193 Parties out of 197 Parties to the UNFCCC have signed the Paris Agreement (United Nations 2021).

12 Onora O'Neill also includes a coherence requirement in her account of the scope of justice and virtue. For details, see O'Neill 1996, especially chapter 4.

13 See, for example, Gewirth 1995; Korsgaard and O'Neill 1996.

14 See, for example, BUND 2019; NABU 2019.

Reference list

Andina, Tiziana. 2018a. 'Transgenerational Actions and Responsibility Toward Our Spe-cies'. PowerPoint Presentation presented at the Our Species and Its Responsibilities. An Ontology for the Environmental Crisis, Turin, Italy. 1 February.

———. 2018b. 'Transgenerational Actions and Responsibility'. *Journal of Critical Real-ism* 17 (4): 364–373. https://doi.org/10.1080/14767430.2018.1488204.

Appunn, Kerstine, and Julian Wettengel. 2019. 'Germany's Climate Action Law'. *Clean Energy Wire*. 18 December. www.cleanenergywire.org/factsheets/germanys-climate-action-law-begins-take-shape.

Association Maurice de Sully. 2019. 'Construction History'. *Notre-Dame de Paris*. www.notredamedeparis.fr/en/la-cathedrale/histoire/historique-de-la-construction/.

Barry, Melissa. 2013. 'Constructivist Practical Reason and Objectivity'. In *Reading Onora O'Neill*, edited by David Archard. London: Routledge.

Bekiempis, Victoria. 2019. 'New York to Pass Most Progressive Climate Crisis Plan in US'. *The Guardian*. 19 June. Online Edition. www.theguardian.com/us-news/2019/jun/19/new-york-climate-crisis-emergency-plan-progressive.

BUND. 2019. 'Die BUND-Bewertung Zum Klimapaket Der Bundesregierung'. *BUND*. www.bund.net/fileadmin/user_upload_bund/publikationen/klimawandel/hintergrundpa pier_bewertung_klimakabinett.pdf.

Bundesministerium für Umwelt, Naturschutz und nukleare Sicherheit (BMU). 2016. 'Kli-maschutzplan 2050'. www.bmu.de/fileadmin/Daten_BMU/Download_PDF/Klimaschutz/klimaschutzplan_2050_bf.pdf.

CDP. 2021. 'Cities A List'. www.cdp.net/en/cities/cities-scores.

Climate Analytics and New Climate Institute. 2021. 'The CAT Thermometer'. *Climate Action Tracker*. November. https://climateactiontracker.org/global/cat-thermometer/.

C40. 2017. '1.5°C: Aligning New York City with the Paris Climate Agreement'. *C40 Knowledge Hub*. September. www.c40knowledgehub.org/s/article/1-5-C-Aligning-New-York-City-with-the-Paris-Climate-Agreement?language=en_US.

———. 2018. 'Barcelona's Climate Action Plan 2018–2030'. *C40 Knowledge Hub*. April. www.c40knowledgehub.org/s/article/Barcelona-s-Climate-Action-Plan-2018-2030?language=en_US.

———. 2020. 'The City of Cape Town's Carbon Neutral 2050 Commitment'. *C40 Knowledge Hub*. July. www.c40knowledgehub.org/s/article/The-City-of-Cape-Town-s-Carbon-Neutral-2050-Commitment?language=en_US.

Deutscher Bundestag. 2019. 'Bundes-Klimaschutzgesetz'. www.gesetze-im-internet.de/ksg/BJNR251310019.html.

Erskine, Toni. 2001. 'Assigning Responsibilities to Institutional Moral Agents: The Case of States and Quasi-States'. *Ethics & International Affairs* 15 (2): 67–85. https://doi.org/10.1111/j.1747-7093.2001.tb00359.x.

———. 2014. 'Coalitions of the Willing and Responsibilities to Protect: Informal Associations, Enhanced Capacities, and Shared Moral Burdens'. *Ethics & International Affairs* 28 (1): 115–145. https://doi.org/10.1017/S0892679414000094.

European Environmental Agency (EEA). 2019. 'Atmospheric Greenhouse Gas Concentrations'. *European Environmental Agency*. March. www.eea.europa.eu/data-and-maps/indicators/atmospheric-greenhouse-gas-concentrations-6/assessment.

Fleming, Sean. 2017. 'Moral Agents and Legal Persons: The Ethics and the Law of State Responsibility'. *International Theory* 9 (3): 466–489. https://doi.org/10.1017/S1752971917000100.

French, Peter A. 1979. 'The Corporation as a Moral Person'. *American Philosophical Quarterly* 16 (3): 207–215.

———. 1984. *Collective and Corporate Responsibility*. New York: Columbia University Press.

Gardiner, Stephen M. 2003. 'The Pure Intergenerational Problem'. *The Monist* 86 (3): 481–500. https://doi.org/10.5840/monist200386328.

———. 2009. 'A Contract on Future Generations?' In *Intergenerational Justice*, edited by Axel Gosseries and Lukas H. Meyer, 77–118. Oxford University Press. www.oxfordscholarship.com/view/10.1093/acprof:oso/9780199282951.001.0001/acprof-9780199282951-chapter-4.

———. 2011a. 'Is No One Responsible for Global Environmental Tragedy? Climate Change as a Challenge to Our Ethical Concepts'. In *The Ethics of Global Climate Change*, edited by Denis G. Arnold, 38–59. Cambridge: Cambridge University Press. www.cambridge.org/core/product/identifier/CBO9780511732294A010/type/book_part.

———. 2011b. *A Perfect Moral Storm*. Oxford: Oxford University Press. https://doi.org/10.1093/acprof:oso/9780195379440.001.0001.

Gewirth, Alan. 1995. *Reason and Morality*. Chicago: University of Chicago Press.

Glasser, Harold. 2011. 'Naess's Deep Ecology: Implications for the Human Prospect and Challenges for the Future'. *Inquiry* 54 (1): 52–77. https://doi.org/10.1080/0020174X.2011.542943.

Goodin, Robert E. 1995. 'The State as a Moral Agent'. In *Utilitarianism as a Public Philosophy*, 28–44. Cambridge Studies in Philosophy and Public Policy. Cambridge: Cambridge University Press.

Hartzell-Nichols, Lauren. 2017. *A Climate of Risk: Precautionary Principles, Catastrophes, and Climate Change*. Environmental Politics/Routledge Research in Environmental Politics 26. New York: Routledge.

IPCC. 2014. *Climate Change 2014: Mitigation of Climate Change. Contribution of Working Group III to the Fifth Assessment Report of the Intergovernmental Panel on Climate Change*. New York: IPCC.

Jamieson, Dale. 1992. 'Ethics, Public Policy, and Global Warming'. *Science, Technology, & Human Values* 17 (2): 139–153. https://doi.org/10.1177/016224399201700201.

———. 2010. 'Climate Change, Responsibility, and Justice'. *Science and Engineering Ethics* 16 (3): 431–445. https://doi.org/10.1007/s11948-009-9174-x.

———. 2013. 'Jack, Jill, and Jane in a Perfect Moral Storm'. *Philosophy and Public Issues* 3 (1): 37–53.

———. 2014. *Reason in a Dark Time: Why the Struggle Against Climate Change Failed – and What It Means for Our Future*. Oxford: Oxford University Press.

Jankovic, Marija, and Kirk Ludwig. 2017. 'Collective Intentionality'. In *The Routledge Companion to Philosophy of Social Science*, edited by Lee C. McIntyre and Alexander Rosenberg, 214–227. London: Routledge.

Korsgaard, Christine M., and Onora O'Neill. 1996. *The Sources of Normativity*. Cambridge: Cambridge University Press.

Kutz, Christopher. 2000. *Complicity: Ethics and Law for a Collective Age*. Cambridge: Cambridge University Press.

Mann, Michael E. 2015. 'The "Fat Tail" of Climate Change Risk'. *Huffpost*. 9 November. https://guce.huffpost.com/copyConsent?sessionId=3_cc-session_c83e09b6-29e5-4161-802b-66b59e2b8852&lang=en-us.

NABU. 2019. 'How Dare You?' *NABU*. www.nabu.de/umwelt-und-ressourcen/klima-und-luft/klimaschutz-deutschland-und-europa/27029.html.

Næss, Arne. 1973. 'The Shallow and the Deep, Long-Range Ecology Movement: A Summary'. *Inquiry* 16: 95–100.

O'Neill, Onora. 1986. 'Who Can Endeavour Peace?' *Canadian Journal of Philosophy Supplementary Volume* 12: 41–73. https://doi.org/10.1080/00455091.1986.10717154.

———. 1996. *Towards Justice and Virtue: A Constructive Account of Practical Reasoning*. Cambridge: Cambridge University Press.

———. 2016. 'Global Justice: Whose Obligations?' In *Justice across Boundaries: Whose Obligations?* Cambridge: Cambridge University Press.

Oppenheimer, M., M. Campos, R. Warren, J. Birkmann, G. Luber, B. O'Neill, and K. Takahashi. 2014. '2014: Emergent Risks and Key Vulnerabilities'. In *Climate Change 2014: Impacts, Adaptation, and Vulnerability. Part A: Global and Sectoral Aspects. Contribution of Working Group II to the Fifth Assessment Report of the Intergovernmental Panel on Climate Change*, edited by C.B. Field, V.R. Barros, D.J. Dokken, K.J. Mach, M.D. Mastrandrea, T.E. Bilir, M. Chatterjee, et al., 1039–1099. Cambridge: Cambridge University Press.

Ornstein, Robert E., and Paul R. Ehrlich. 1989. *New World New Mind: Moving Toward Conscious Evolution*. Cambridge, MA: Malor Books.

Page, Edward A. 2006. *Climate Change, Justice and Future Generations*. Cheltenham: Elgar.

Shankman, Sabrina. 2019. 'Coasts Should Plan for 6.5 Feet Sea Level Rise by 2100 as Precaution, Experts Say'. *Inside Climate News*. 21 May. https://insideclimatenews.org/news/21052019/antarctica-greenland-ice-sheet-melting-sea-level-rise-risk-climate-change-polar-scientists.

Skinner, Quentin. 2009. 'A Genealogy of the Modern State'. In *Proceedings of the British Academy, Volume 162*, edited by Ron Johnston. Oxford: Oxford University Press.

Storrow, Benjamin. 2019. 'New York Has a Climate Plan – Now It Has to Follow Through'. *E&E News*. 29 June. Online Edition. www.scientificamerican.com/article/new-york-has-a-climate-plan-now-it-has-to-follow-through/.

Taibi, Fatima-Zahra, and Susanne Konrad. 2018. 'Pocket Guide to NDCs Under the UNFCCC'. https://orbit.dtu.dk/en/publications/pocket-guide-to-ndcs-under-the-unfccc.

Thompson, Janna. 2002. *Taking Responsibility for the Past: Reparation and Historical Injustice*. Cambridge: Polity.

United Nations. 2021. 'Paris Agreement'. *United Nations Climate Action*. February. www.un.org/en/climatechange/paris-agreement.

United Nations Climate Change. 2015. 'Paris Agreement to the United Nations Framework Convention on Climate Change'. https://unfccc.int/sites/default/files/english_paris_agreement.pdf.

United Nations Environment Programme (UNEP). 2019. *The Emissions Gap Report 2019*. Nairobi: UNEP.

Valentini, Laura. 2011. *Justice in a Globalized World: A Normative Framework*. Oxford: Oxford University Press.

Wehrmann, Benjamin. 2019. 'Germany's 2030 Climate Action Package'. *Clean Energy Wire*. 14 October. www.cleanenergywire.org/factsheets/germanys-2030-climate-action-package.

White, Lynn Jr. 1967. 'The Historical Roots of Our Ecologic Crisis'. *Science* 155 (3767): 1203–1207. https://doi.org/10.1126/science.155.3767.1203.

4   Including future people for their own sake

The moral value of future persons

In Chapters 2 and 3 I argued that if an account of the scope of climate justice is to be successful, it must be able to show that we have intergenerational obligations despite the uncertainty and complexity of climate change. I showed that we have reasons to include future people within the scope of our obligations despite the challenges of uncertainty and complexity and, more specifically, that we have an obligation to respect the agency and autonomy of future people whenever we engage in intergenerational actions. This chapter will focus on a third important question, namely how future people ought to be included within the scope of our obligations. This final question concerns the very basis of an account of intergenerational justice, namely how we ought to think about future people when we think about justice and its scope. I will argue that to fully overcome the intergenerational climate challenge, an account of scope must meet the following criterion C3:

> *INCLUDING FUTURE PEOPLE FOR THEIR OWN SAKE. An account of the scope of climate justice must include each person for their own sake and as an end in themselves.*

It is important to determine the way in which future people are to be included in the scope of justice before we can move on to develop a substantive account of climate justice. Doing so complements criteria C1 and C2 by completing our account of scope and simultaneously setting the outer boundaries for and determining the ground rules of possible substantive accounts of justice. This ensures our account of scope can provide a sound basis for principles of intergenerational justice. In previous chapters, I argued that we have an obligation to respect the autonomy conditions of those others whom our actions presuppose; I will now argue that unless we include future people in the scope of our obligations for their own sake, we risk not being able to fulfil this basic requirement of justice. I will argue that unless our account complies with C3, it will sanction using future people merely as means to our ends. We risk using future people as mere means, for example, whenever we fail to take on our share of the mitigation burden. The requirement to include future people for their own sake thus enables an account of scope to fully account for the

DOI: 10.4324/9781003258902-4

special moral wrong of behaviours – like intergenerational buck-passing – that we often exhibit in actions relevant to climate justice. Being able to explain these moral wrongs ensures the normative accuracy of the account and its ability to overcome all moral challenges posed by intergenerational climate change. To motivate the criterion, this chapter will consider the possible implications of failing to abide by it. I will argue that these consequences are unacceptable but can be avoided if we construct the account in such a way that it includes each person, including future people, for their own sake. The alternative scenario I am going to rely on is drawn from Samuel Scheffler's account of our relation to the future.

Scheffler draws a striking picture of our relationship to future generations. One of his central claims is that past, present, and future generations constitute the "ongoing, historical project" of humanity, a framework we need in order to be able to value many of the things we currently value. Because we are so reliant on this framing by an ongoing human history, we become dependent on future generations – in the sense that we need them, the "collective afterlife," to exist – in order for most of what we care about to continue mattering to us.[1] In turn, Scheffler believes our dependence on future people gives us reasons to attend to their interests (Scheffler 2013).

Scheffler's account draws on P.D. James' novel *The Children of Men*, in which James envisages a world where widespread infertility threatens the imminent extinction of humanity (call this the *infertility scenario*). Scheffler suggests that, when faced with an infertility scenario, "people would lose confidence in the value of many sorts of activities, would cease to see reason to engage in many familiar sorts of pursuits, and would become emotionally detached from many of those activities and pursuits" (Scheffler 2013: 44). This is his "afterlife conjecture."

He believes most people would react to such a scenario with "apathy, anomie, and despair" and a "pervasive loss of conviction about the value or point of many activities." Societies would face "the erosion of social institutions and social solidarity" as well as "the deterioration of the physical environment" (Scheffler 2013: 40). The upshot is that, when faced with the imminent extinction of humanity, many of our valued projects and activities would become less important to us. Scheffler's conjecture reveals both the temporal dimension of valuing and what he calls our conservatism about value. In his view, part of what it means to value something is wanting to preserve and sustain it over time, and we need humanity to have a future for this to be possible.

It also suggests that the survival of humanity matters to us beyond being a vehicle for carrying on *specific* valued projects. In addition, he believes human life as a "thriving, ongoing enterprise" is an important element of the background against which we develop a sense of what is valuable and make judgements of importance. It provides a frame of reference without which "our sense of importance . . . is destabilized and begins to erode" (Scheffler 2013: 60). Consequently, though we would not lose our ability to make value judgements altogether, in an infertility scenario "our ability to apply the concept [of valuing] confidently and unreflectively would be compromised and . . . the range of activities that seemed to us worthwhile would be drastically diminished" (Scheffler 2013: 183).

In addition to future-oriented projects – such as cancer research – whose ultimate success can only be expected sometime after our death, Scheffler suggests that a much wider range of projects would likely lose their appeal without the prospect of a collective afterlife (Scheffler 2013).[2] People may, for example, be less inclined to pursue artistic or literary projects or, perhaps even more so, to procreate and engage in the wide range of activities related to parenting (Scheffler 2013: 26). The afterlife conjecture thus seems to show that, in addition to mattering to us in its own right, the collective afterlife is an important condition of many things mattering to us now, in the sense that what matters to us depends on our confidence in the existence of a collective afterlife (Scheffler 2013: 51).[3]

Scheffler also notes that, remarkably, we are not usually as demoralised by the prospect of our own death. A great number of individuals who do not believe in a personal afterlife after their biological death live fulfilled and value-laden lives knowing that one day they will simply cease to be (Scheffler 2013: 71–73). This leads him to claim that in some respects – including as a condition of things mattering to us – the survival of future generations "matters more to us than our own." The fact that future people matter so much to us, to the extent that we are dependent on them if we are to live value-laden lives, gives us reasons to attend to their interests (Scheffler 2013: 78).

Though we may rarely stop and think about the extent to which we implicitly rely on the existence of an afterlife, the degree of dependence Scheffler describes seems plausible. I am therefore going to assume that he is right in believing that the prospect of imminent extinction would cause a general sense of despair. I am also going to assume that his corollary claims – that the reason for this is that we care about future generations as a condition of other things mattering to us as well as in their own right – are accurate. But while I find the descriptive element of his account plausible, I believe it has undesirable implications for intergenerational justice. Perhaps Scheffler did not intend for his work to be interpreted this way, but I believe it is plausible to understand his account as follows: our dependence on future generations justifies an obligation to attend to their interests, including as part of a theory of intergenerational justice. In other words, future generations are to be included within the scope of our obligations because of their derivative importance to us as a condition of other things mattering to us.

It strikes me that, because of its implications for an account of justice, this kind of interpretation ought to be avoided. C3 serves as a way to sidestep it. Let me emphasise that Scheffler's goal is not to give either a full account of intergenerational justice or an exhaustive explanation of our motivations to care for the interests of future people. My arguments in this chapter should therefore not be read as objections to Scheffler's account but as attempts to complement it. Scheffler's project differs from mine in that he focuses on describing the extent to which generations are already interconnected rather than laying out the normative foundations for intergenerational justice. Precisely because he stops at these limited conclusions, it will be interesting to unpack his account and its possible implications as part of a normative discussion of intergenerational climate justice.

The ways in which Scheffler's account relates to my approach to scope rest on an important distinction between two distinct levels of moral theorising, namely *justifying* moral rules and *grounding* their application. I understand the process of justification, on the one hand, as that which explains why the chosen standard of moral rightness is indeed the correct standard of rightness. With regard to my proposed account of scope, the relevant standard of moral rightness is the inclusion of future people within the scope of climate justice. The appropriate justification for including future people within the scope of climate justice, as argued in Chapters 2 and 3, is their entitlement to the conditions of autonomy and respect for agency. This justification is tied to the way in which future people are included in the scope of our obligations: when future people are included in the scope of justice for their own sake, what justifies their inclusion is their entitlement, as ends in themselves, to the conditions of autonomy. Conversely, if future generations were to be included in the scope of our obligations based on their importance to us and our goals, what would justify their inclusion would be their standing as means to our ends. I have argued for this before and will again, in this chapter, argue that we ought to justify our intergenerational obligations not in reference to our dependence on the future but based on the equal moral value of future people as ends in themselves and hence that we ought to include future people in the scope of justice for their own sake.

The process of grounding our obligations, on the other hand, takes place on a distinct level of moral theorising. This level describes the context in which, in a particular case, an obligation is triggered under the specified standard of moral rightness. It explains how the overarching moral rule is to be applied to the context at hand. This can be done, for example, by a process of deliberation or in reference to the agents' actions or interconnections. On the account of scope I propose in this book, the work of grounding our obligations is done by the presuppositions about future people that underlie our intergenerational actions. These presuppositions reveal the connections between present and future agents and, for the agents who engage in them, trigger obligations of justice to those whom their actions presuppose.[4]

In Chapter 5, I will argue that Scheffler's account has an important role to play in grounding obligations to future people, for it shows that many of our most common activities are in fact intergenerational actions. The reliance of present generations on the future that is brought to light by his account thus has important implications for what we owe future people by triggering intergenerational obligations in the specific context of our actions. First, however, in what follows I am going to argue that a successful account of intergenerational climate justice must provide a justification for these obligations that is based on the equal moral value of future people as ends in themselves and hence include future people in the scope of justice for their own sake rather than as means to satisfy our interests.

I will begin in the next section by arguing that, if we were to include future people in the scope of justice based on their importance to us and fail to justify our obligations to them with their entitlement to the conditions of autonomy, we would risk wrongly using future persons as mere means to satisfy our own interests. Most importantly, failing to include future generations as ends in themselves may sanction unjust future worlds that are clearly incompatible with the interests of future persons. In the following section I will defend my arguments against an important objection, which holds that on Scheffler's account future generations are not an instrument in the conventional sense of the term. I will then consider Scheffler's claim that part of our dependence on the afterlife comes from the fact that we also care about it in its own right. I will argue, first, that there are empirical grounds to doubt this claim and, second, that care is not an appropriate justification for duties of justice.

Ends and means

My first argument concerns how, as part of an account of scope, we ought to interpret the basic requirement of justice that each person is of equal moral value. On any understanding of justice in which agency and autonomy are key, respecting persons' equal moral value must entail a duty to respect their freedom to choose and pursue their own ends. In turn, it must prohibit using others as mere means to the agent's ends. In what follows, I argue that unless an account of the scope of climate justice includes future people for their own sake, we risk wrongly using them as mere means to our ends and thus fail to meet one of the basic tenets of justice. In other words, part of what it means for an account of scope to respect persons' equal moral value is to guard against using others as mere means. Unless we complement our account with the requirement to include future people as ends in themselves and independently of their derivative importance to us, our account may sanction using future people as mere means in morally objectionable ways.

A similar concern has been expressed before. In his commentary on Scheffler's lectures, Niko Kolodny writes that on Scheffler's account the existence of future people

> is a necessary condition of many objects of broadly egoistic concern. . . . Although it isn't quite right, one might try to bring out the egoism by saying that we value future people as "means." It's just that the afterlife conjecture shows that future people are much more essential means than we might have thought.
>
> (Kolodny 2013: 173)

Kolodny seems hesitant to flesh out this point ("Although it isn't quite right") and stops short of developing his comment into a full argument. Perhaps, his hesitation stems from a general suspicion that we may not be able to treat future, non-existent person as mere means in the way this is traditionally understood. I believe that such a suspicion would be misplaced.

One way of understanding what treating someone as a mere means entails links it to a person's ability to consent to the end in question. Thus, Onora O'Neill holds that "[a]n agent treats another merely as a means and thus wrongly if in his treatment of the other the agent does something to which the other cannot consent" (O'Neill 1985, cited in Kerstein 2009: 172).[5] Her statement contains two crucial points. The first is that others are treated as means if they are denied the possibility to consent to the end in question. The second point to take from O'Neill's claim is that it is only wrong to use others as means if they are used *merely* as means to the agent's end. That suggests that it may not be wrong for others to serve as a means to the agent's end if they are also thereby treated and respected as ends in themselves. O'Neill gives only an abstract account of what this requires: to treat others as ends in themselves, we must respect them as persons by taking into account and respecting their particular capacities for rational and autonomous action (O'Neill 1985: 264). One plausible interpretation of what this may require in practice is that our actions must treat the other person as an autonomous agent by respecting their own interests and conditions of autonomy.

It is easy to think of examples in the intragenerational context that fail to meet this requirement. For example, a mugging at gunpoint is an action which clearly does not meet the requirement to also treat the other as an end. In this case, the mugger coerces the victim to bend to his will by forcing her to hand out her wallet. The victim cannot consent to being used in such way as she is threatened with severe harm and thus has no choice but to acquiesce. She thereby serves as a means to the mugger's end of robbing her, an end which clearly violates her autonomy.

Acts of the same structure can occur between non-overlapping present and future generations, too.[6] Environmental degradation is one particularly striking example of this. By changing and degrading the natural environment, we are increasingly able to influence the future in severe and irreversible ways.[7] Past and present generations have caused and are still causing changes in the earth's climate that will continue affecting the earth for centuries (IPCC 2014a, 2021). Some changes caused by past greenhouse gas emissions have already manifested, and some lagged effects of past emissions are already unavoidable. Nevertheless, continued emissions are increasing the magnitude of future changes and consequently the danger of harm to future generations. The actions of current polluters, in other words, threaten to cause severe harm to future generations (cf. IPCC 2014a, 2014b, 2018, 2021). In light of this threat, consider the current mitigation efforts on a global level. On the one hand, Parties to the Paris Agreement, which includes most nation-states, have committed to the goal of keeping global average temperature rise to a maximum of 2°C above pre-industrial levels (United Nations Climate Change 2015).[8] On the other hand, the mitigation efforts that Parties are actually pursuing are only enough to limit warming to approximately 2.7°C (Climate Analytics and New Climate Institute 2021).

A highly simplified but helpful way to conceptualise these numbers is in terms of a carbon budget shared between generations. We can think of the amount of emissions predicted to cause an average temperature rise of 2°C as the available carbon budget. A fair division of the remaining budget would intuitively seem to entitle each generation to a certain share.[9] On any plausible interpretation of what a fair share is, present generations are using more than what they are entitled to.[10] This is suggested by the fact that current policies are failing by a large margin to meet the mitigation target. If current policies, including pledged policies, are continued, emissions would greatly exceed the carbon budget. To avoid this and the connected environmental harms, future generations will have to take up a much larger share of costly mitigation action. These observations suggest that present generations – represented, in the Agreement, by their states' current governments – are pursuing the following end: to avoid costly mitigation efforts by continuing to emit more than their fair share, while still achieving (or not thwarting the possibility of achieving) the 2°C goal they themselves have set in a legally binding agreement.

One can now see how, by passing the mitigation buck to future generations, present generations use future people as mere means. The only way present generations can achieve their end is if future generations use considerably less of their own carbon budget or, alternatively, develop negative emission technologies necessary to counteract past emissions.[11] The climate changes to which present actions are contributing pose a threat of severe harm to future generations, which will prevent many future people from fulfilling their basic needs. This means they will not be able to choose freely whether they want to take up the slack of past generations through greater mitigation efforts themselves. Unless they do, they will increasingly be feeling the harmful effects of climate change. Recalling what was said earlier, we may therefore say that in passing the buck to future generations, future people are used as mere means by present generations. To avoid climate harms, future generations have no choice but to take up the role envisaged for them by present agents in the pursuit of their interests. At the same time, passing the buck to future people is an action which clearly does not respect future people as ends in themselves. By forcing future generations to choose between suffering climate harms and contributing to present agents' end, it is reminiscent of our earlier example of a mugging at gunpoint and a clear violation of future people's conditions of autonomy. Moreover, the goal for which future generations serve as means is so far removed from what we may reasonably presume their interests to be that we cannot assume it to be an end they would share. On any plausible account of what we can assume future people would consent to, being forced to take up our share of the mitigation burden that we were unwilling to shoulder is almost certainly not part of it. By failing to take up a large enough share of the mitigation burden, present generations thus fail to respect future people as ends in themselves and instead treat them as mere means. This seems to constitute a special moral failing by present generations. If our account of scope is to be normatively accurate, that is, account for all factors that are morally relevant to an account of scope, it is important that it be able to account for this.

A problem of the same structure applies to Scheffler's conjecture, too. According to Scheffler, we need future people to achieve a very specific and very important end. We need confidence in the afterlife for many other objects and activities to matter to us now, but we also "need humanity to have a future for the very idea that things matter to retain a secure place in our conceptual repertoire" (Scheffler 2013: 60). What Scheffler suggests is that the existence of future people is a necessary means for us to achieve our goal: we need to be confident that future people will exist if we are to retain our ability to confidently value things.

Although it is true that future people cannot consent to serving this end, it is an end which is in their interest, too, and which is not in violation of their agency and entitlement to autonomy. In fact, it is an end we may plausibly assume future people would consent to, as present and future generations alike are likely to share the interest in upholding the "ongoing enterprise" of humanity. It therefore does not seem objectionable for future people to serve as a means to secure the existence of an afterlife and enable the practice of valuing. For the moment, then, we can say this: though future people are a means to the continuation of humanity and, in turn, to our being able to value things, this in itself does not amount to treating them merely as means and thus wrongly, for both are goals that are plausibly in their interest, too, and not in violation of their autonomy.

Yet the main problem posed by an account of scope based on the derivative importance of future people lies in the behaviours that such an account would sanction. An account of scope that justifies our obligations based on the derivative importance of future people to us, and in which future people are therefore included as means to the continuation of the afterlife rather than for their own sake, pays insufficient attention to the interests of future persons. This, in turn, has unacceptable implications for intergenerational justice. An account of justice which does not consider the interests of future people for their own sake is less likely to limit the future-affecting ends present people may pursue to those that are plausibly in the interest of future people and that respect their autonomy. In other words, relying on present persons' interest in the afterlife as a justification of intergenerational justice sanctions unjust future worlds as it underdetermines what kind of future we should strive for to continue the "ongoing enterprise" of humanity.

The upshot is, first, that such an account would fail to sufficiently protect future people from being wrongly used as mere means for subordinate ends which are not in their interests and which contribute to creating unjust futures. Moreover, in addition to failing to prevent the special moral wrong of using others as mere means, the resulting account would offer no protection against the vulnerability future people face by virtue of their position in time – one of the main issues identified by the intergenerational climate challenge. Finally, failing to address these concerns through sound moral norms will leave present generations particularly susceptible to the kind of moral corruption identified by Gardiner. That is, without a set of norms that protect future people and their interests for their own sake, present generations may be more likely to distort existing moral norms in ways that promote their own interests over those of future generations. Conversely, the requirement to include future people in the scope of justice for their own sake would, on a substantive level, compel us to take their foreseeable interests into consideration in the pursuit of all intergenerational activities. This would successfully guard against building unjust futures and using vulnerable future people as mere means to do so.

As an example, consider the case of the present agent Peter. Peter owns various companies that, among others, emit large amounts of greenhouse gases and contribute to environmental degradation, climate change, and the negative effects on the human and natural environment associated with both. Nevertheless, his business matters a lot to him. According to the afterlife conjecture, Peter needs humanity to have a future for his projects, including his business, to continue to be of value to him. More broadly, he also needs humanity to have a future for the practice of valuing to continue making sense to him and his life to continue being a value-laden, human life. An account of intergenerational justice that follows the afterlife conjecture and includes future people because of their derivative importance to present generations would require Peter to do what is in his power to avoid human extinction. This, however, seems to be nearly everything it would demand. As long as he can be confident of the survival of a sufficiently large group of future persons with the ability to preserve and sustain what he values, Peter's personal interest in the afterlife will have been fulfilled. These future persons provide the framework Peter needs to be able to value things now, including his fossil fuel business.

An account that includes future people not for their own sake but as means to our ends is more likely to be silent about what kind of future Peter, and present generations in general, should strive to create for humanity. It need not say anything, for example, about the potential of Peter's projects to further the economic dependence on fossil fuels, increase future inequalities, worsen the living conditions of many future persons, or degrade the natural environment in ways that would limit the conditions of autonomy of future people. What would likely be the principal concern of such an account is not what possible afterlives may look like, but the ability of the afterlife to serve as a projection for present persons' projects and values and as a vehicle to enable the practice of valuing itself. This presupposes future people with a general capability to value projects, objects, and practices and those who are capable of understanding, sharing, or developing our values. Yet it does not require present generations to strive for a future in which other interests of future people, including fundamental interests in the conditions of autonomy, are met. Such an account sanctions future worlds in which humanity continues to exist but under unjust circumstances. In turn, such an account would allow present generations to treat future people as mere means to their unjust projects. In Peter's case, future people are a means to his ends insofar as his confidence in their existence is a necessary condition for his ability to value his projects and practices. Yet Peter's pursuits will, among other things, affect the environment in ways that will be increasingly hostile to many basic human interests and prevent many future people from enjoying the conditions necessary to live an autonomous life. Future generations are thereby a means to upholding the value of practices that are presumably not in their interests and are likely to violate many of their conditions of autonomy. Thus, by sanctioning unjust futures, an account that includes future people because of their derivative importance to us is also going to allow wrongly treating future people as mere means.

Note, again, that Scheffler's arguments are not intended to provide a full account of intergenerational justice, nor is my claim that in addition to the survival of humanity we must aim for a just future necessarily inconsistent with his view. Accordingly, what I say should not be read as an objection to Scheffler's arguments. My aim throughout this chapter is to show, in reference to his conjecture, that justifying obligations of intergenerational justice based on the derivative importance of future people has unappealing consequences we should try to avoid by making our account of scope consistent with a criterion like C3. In other words, while Scheffler's account does not go far enough to secure justice for future generation, it can help us develop a successful account of scope by pointing us in the direction of what is missing.

A game analogy

A possible objection to what I just argued is that drawing such a sharp distinction between the afterlife as a condition of things mattering to us and as something that matters to us in its own right misrepresents the role of the afterlife within Scheffler's account. As Scheffler describes it, the afterlife is not a mere instrument which enables valuing but something which partly constitutes the community of human valuing. The way in which future generations serve as a means to, or are a condition of, things mattering to us is as common participants in the "ongoing enterprise" of humanity (Scheffler 2013: 60). What he describes is not a case of present generations one-sidedly instrumentalising future persons. Instead, the afterlife enables us to value things by partly constituting, together with ourselves, a common and cooperative project, thereby providing the necessary framework for our judgements of importance. Like future generations, present persons too are a necessary means to this common enterprise, leading to a state of interdependence rather than the one-sided use of one generation by the other.

This situation of mutual dependence resembles that of a game. Imagine, for instance, a game of chess. Here, each player's participation partly constitutes the game as well as being a means to the other player's ability to play. Moreover, assuming each player seeks to win, both players are means to an end that goes against their own interests. It is unreasonable to believe, and we usually do not, that players in game settings wrongly use one another as mere means. If the analogy holds, it follows that it would be equally misguided to accuse present generations of using the afterlife as a mere means, even if it is in pursuit of an end – such as Peter's ability to value his environmentally destructive pursuits – that is presumably not in their interest. Future generations, too, will in turn be able to rely on previous and successive generations as means to valuing their own, possibly equally destructive, pursuits.

The first point to note in response is that it is possible for something that constitutes an end to simultaneously be a means to it. Parfit, for instance, notes that the relation between means and ends does not always have to be causal (Parfit 2001: 20). As an example, he considers the case of the second-born son to a king: the death of his elder brother constitutes his succession to the throne rather than causing it. At the same time, his brother's death is a necessary means to his end of becoming king (Parfit 2001). Consider, also, a ruler wanting to show off his power through a public execution. The prisoner's death does not cause his display of power but constitutes it. In addition, the ruler treats the prisoner merely as a means and not as an end in himself, as his actions clearly show no respect for the prisoner's autonomy and personhood. The fact that something fully or party constitutes an end does not imply that it cannot simultaneously be a means, or even merely a means, to it. This, then, is not what gives bite to the objection, nor the reason why we do not think players in a game wrongly use one another as mere means.

The second, more important point relates to what it takes to be used merely as a means and thus wrongly. Recall that an agent only treats another as a mere means if in his actions he also does not respect the other as an end in herself, for example, by allowing her to consent or dissent to the action in question. While players in competitive games serve as means to their competitors' ends, they have freely consented to being a part of the game. Neither player is forced by the other to adopt her competitor's end. They have each consented to being used in the ways stipulated by the rules of the game while in full possession of their agency and autonomy. In this case, the players' consent to the game and its rules ensures they are also treated as ends and therefore not wrongly (see also Kerstein 2009: 173).

Clearly, the case of future generations differs. Unlike competitive players, future persons cannot give their actual consent to how they are being treated by present generations. Peter, for example, cannot seek future generations' consent to serving as a means for his goal of valuing his fossil fuel business. By virtue of their position in time, future generations cannot provide their actual consent to any of our present projects and pursuits. This suggests the following: whenever our intergenerational actions do, in the ways described earlier, rely on treating future generations in a certain way, it is crucial that we instead seek to respect future persons as ends by respecting their autonomy and plausible future interests.

Fleshing out the details of what this is going to require is a question that is best answered by a substantive account of intergenerational justice. However, by stipulating that future people must be included for their own sake, an account of the scope of justice can set the necessary outer boundaries within which adequate principles of justice can later be developed. Stipulating that future people are to be included in the scope of justice as ends in themselves and not for their derivative importance to present persons ensures that future interests and conditions of autonomy, too, are protected for their own sake. This ensures future entitlements cannot, without justification, be sidelined in the pursuit of present persons' ends. In this way, how we include future people within the scope of our obligations sets the ground rules for intergenerational justice, shaping the content of the substantive principles our account allows. This, in turn, determines whether our account can successfully serve as a tool to meet the challenges of intergenerational climate change in a just manner.

What I have argued is that, even in situations of mutual dependence, if we fail to include future people for their own sake we might not take sufficient account of their interests and risk using them as mere means. This increases the danger of moral corruption and fails to account for the vulnerability of future people identified as part of the intergenerational climate challenge. Instead, because future people cannot consent to our projects, it is important that our actions treat them as ends by respecting their interests and autonomy. Including future people for their own sake – and thus, in turn, considering their interests and entitlements for their own sake – is one way in which our account of scope can ensure this will be the case.

The importance to us of the afterlife in its own right

I have so far focused on one part of Scheffler's claim, which holds that the afterlife matters to us as a necessary condition of other things mattering to us. I argued that, rather than because of their derivative importance highlighted by Scheffler's argument, future people must be included in the scope of justice for their own sake. In other words, although our obligations to future people can be grounded in the presuppositions of our actions, what justifies them must be the equal moral value of future persons. Our account must include future people for their own sake in order to ensure this. Yet Scheffler notes that the afterlife also matters to us in a second way, namely in its own right. This sets the afterlife apart from other necessary conditions for human life, such as the provision of oxygen: the afterlife is not just a causal condition of other things mattering to us but the reason for their doing so (Scheffler 2013: 26). It is in part because we already care about the survival of humanity in its own right that Scheffler believes an infertile world would prompt apathy and despair.

These two ways in which the afterlife matters according to the conjecture cannot easily be disentangled so that drawing such a stark distinction between them may seem artificial. I already considered an objection to this extent earlier in the chapter. My focus in this chapter, however, is not on a detailed examination of Scheffler's account but on how we may use his arguments as a starting point for constructing a successful account of the scope of climate justice. In order to do so, it is important to assess each claim in its own right. Most of what Scheffler argues highlights the ways in which what matters to us is dependent on the afterlife: the existence of the afterlife allows us to personalise our relation to the future, conserve our values, and uphold the circumstances in which valuing as we know it is possible (see, for instance, Scheffler 2013: 29–30, 51, 59–61). In previous sections, I relied on this first part of Scheffler's claim to argue that an account of scope may not justify our obligations to future people based on their derivative importance to us but that it must instead value and include them for their own sake. In what follows I will argue that Scheffler's second claim – that the afterlife also matters to us in its own right – shows that an account of scope cannot justify our obligations based on our care for future people either. There are two reasons for this. On the one hand it is unclear that, beyond the mere existence of future people, we also care about their interests being met; on the other hand, the notion of care is itself not an appropriate justification for justice.[12]

How much do we really care about the afterlife?

Scheffler motivates the claim that we care about the afterlife in its own right by noting our willingness to pursue projects whose benefits will only be reaped by future generations. An example of this is present generations' engagement in researching a cure for cancer. His argument can be summarised as follows:

P.1   We are instrumentally rational; that is, we work on projects to achieve a certain goal and only if said goal is achievable.
P.2   We are happy to pursue projects whose expected benefits will only manifest after our death.
P.3   The goal of these projects is to achieve their expected, future benefit.

C.1   We pursue these projects to benefit future generations (call these *future-oriented projects*).
C.2   Future generations matter to us in their own right (see, for example, Scheffler 2013: 26–27).

Taking Scheffler's second claim – that we care about the afterlife in its own right – as the reason to include future people within the scope of justice would have very different implications for the resulting account of scope and how future people are included within it than what I described earlier. It would likely imply that we have reasons to want humanity to have not just any future but a certain type of future, that is, a future in which the interests of future persons are met. Some of what Scheffler says seems to support this stronger claim. For instance, he writes that we "have reasons . . . for attending to the interests of future generations" and the means "to promote the survival and flourishing of humanity after our deaths" (Scheffler 2013: 78).

Insofar as future generations' mattering to us in their own right is understood not as their mere existence but as their well-being and interests mattering to us, an account of scope based on this will take account of and require us to respect the interests of future people. Indeed, thus understood our care for future people would seem to result in an account of scope that includes future people in what I argued is the correct way, that is, for their own sake, and which in turn does not fall victim to any of the objections raised in earlier sections. Yet it is unclear that our current future-oriented activities can support this stronger claim about care and, in turn, that any resulting account of scope could successfully be applied to the current problem background. There are two reasons to doubt this. On the one hand, Scheffler's stronger claim relies on speculations about the reasons people have to engage in future-oriented projects and in doing so obscures possible alternative motivations. On the other hand, present generations engage in a number of actions that are pushing in the opposite direction by creating a more harmful world for future generations. This suggests that we may not in fact care about the future as much as Scheffler believes.

As a placeholder for future-oriented projects of the first kind, consider Scheffler's own example, people' willingness to engage in cancer research. With regard to cancer research, P. 3 holds that researchers' goal in undertaking such work is to benefit future generations. I believe this claim can be disputed. While it may be true for some, it is far from obvious that all researchers are only or even primarily motivated by a payoff that will occur after their deaths. Nor need this be the case for the action to be classed as an intergenerational action. Recall that I considered an objection to this extent in Chapter 3. What, I argued, makes an action intergenerational is that it is conceptually reliant on the presupposition that there will be future people. Long-term medical research which we know is not going to bear fruits within our lifetime seems to fall under the category of strong intergenerational actions: it presupposes future people who will be able to reap its benefits, as well as future people who will be able and willing to contribute to it. This act description, however, does not determine individual motivations for engaging in the intergenerational action. In the case of cancer research, for example, it is equally plausible for individual researchers to be motivated by some payoff to be received within their lifetimes, be it the wish to take on the challenge of solving a complex medical problem (see Frankfurt 2013: 134), scientific curiosity, or fame and appreciation by colleagues and the public. These goals are not in competition with the intergenerational character of the action but goals that individual agents hope to reach by engaging in that intergenerational activity. In other words, a medical problem that is not so complex as to require multiple generations of researcher to participate in solving it may not, for example, be seen as an equally interesting challenge to take on or draw the amount of attention and fame that is sought. Similar things as for cancer research can be said about agents' motivations to engage in other future-oriented actions, including trying to prevent severe climate change or environmental degradation. Yet while agents' alternative motivations do not detract from the intergenerational character of the action, they do question the applicability of Scheffler's argument to many future-oriented actions. That we can think of many plausible reasons to engage in such actions beyond care for the afterlife at least questions the extent to which engaging in them shows that future generations matter to us in their own right.

Let's nevertheless, for the sake of the argument, assume for a moment that agents who engage in future-oriented projects do this because they care about the afterlife in its own right. Even if this were to be the case, their activities are in competition with a number of actions that, instead of benefitting future generations at the cost of present people, defer the costs of present benefits to future generations and actively worsen future living conditions. If actions whose benefits occur after the agents' deaths suggest their agents care about the afterlife, these actions must in turn suggest their agents' lack of care and a stark disinterest for future interests. An obvious example is the failure of present and past generations to take meaningful action against climate change and their continued engagement in highly emitting or otherwise climate changing activities. Despite the global awareness of the dangers these actions pose to future generations, many nations are continuing to endorse and even encourage activities like deforestation or the burning of coal (Food and Agriculture Organization of the United Nations 2020; Driskell Tate et al. 2021). While on the global level the problem of climate change is compounded by a coordination problem between actors that inhibits large-scale action, single agents, including the relevant collective agents, have nevertheless long been able to take individual action to show either their commitment to the cause or lack thereof. For instance, the decision by Donald Trump's administration to remove the US from the Paris Agreement, though opposed by a large part of the population and fortunately reversed by his successor Joe Biden, could suggest that a significant number of US citizens do not care, or not enough, about the afterlife to want to protect it from the possibility of dangerous climate change. Similarly, if devoting one's career to cancer research shows that one cares about the afterlife, then working for or investing in the fossil fuel sector, as many currently do, must suggest that people either do not care at all or at least not enough for this care to win out against other concerns.

This is not to say that no present people care about the afterlife in its own right, or that the majority does not. To the contrary, it is plausible that many of us do care about future generations. It does, however, suggest that it is likely that a significant number do not, or not sufficiently to have reasons to protect future people's interests. This casts enough doubt on the claim that we, as a present generation, care about the afterlife in its own right to make it at best a very shaky foundation for an account of intergenerational justice. An account of the scope of climate justice, however, should provide a sound basis for action to all agents in order to meet the challenges of intergenerational climate change.

Care and the justification of justice

The second reason we ought to look for an alternative justification for intergenerational obligations is that the fact that others matter to us – in the sense that we feel concern or affection towards them, in other words, that we care about them – is not an appropriate reason to extend considerations of justice to them. This is closely connected to what I just argued. The previous argument showed that care for the afterlife provides at best a patchy coverage of present agents. While some agents are likely to care, others presumably do not, and we have no reliable indication of whether one does. Because it is of such personal nature, the notion of care is not an appropriate reason to extend considerations of justice to others.

The purpose of principles of justice is to provide impartial rules to protect people's fundamental needs and adjudicate between conflicting claims on the necessary material or non-material goods. Though it does lead to including future people for their own sake, an account of justice which takes a contingent personal preference such as care as the justification for including others within its scope does not seem able to mediate fairly and impartially between persons. The resulting principles would therefore not be action-guiding to agents who do not meet its justificatory requirement, that is, who do not care about future generations. In order to be impartial and action-guiding for all agents, principles of justice ought to be neutral with regard to both their subjects and duty-bearers. Such principles can then provide moral rules that are applicable to all agents regardless of individual motivation. If our account of scope is to be action-guiding for all agents, as I argued it needs to be in order to overcome the intergenerational climate challenge, it must not only include future people for their own sake but also apply equally to all, independently of contingent personal preferences. Both requirements are met by including future people within the scope of justice as ends in themselves and based on their entitlement to agency and autonomy.

Conclusion

In this chapter, I argued that unless an account of the scope of climate justice is justified by future people's equal entitlements to the conditions of autonomy and includes future people as ends in themselves it will sanction using future people as mere means and allow for unjust future worlds. We can guard against this by adopting a criterion like C3, which requires that an account of scope include future people for their own sake. Fixing this as part of our account of scope sets important ground rules for any substantive principles of climate justice that may build on the account, ensuring they respect the autonomy conditions of both present and future people. This allows us to meet the basic requirements of justice and overcome the intergenerational challenge in a just manner.

More specifically, complying with C3 allows the account to be normatively accurate and action-guiding in the face of the intergenerational challenge. On the one hand, it allows the account to pick up instances of using future people as mere means which are relevant to the moral evaluation of our actions and in turn to provide normatively accurate guidance. Passing the buck of climate action to future people is an instance of treating future people as mere means, and it is important that an account of scope be able to fully account for this special moral wrong in order to successfully meet the intergenerational challenge. On the other hand, including future people for their own sake and based on their entitlement to agency and autonomy allows the account to be action-guiding for all agents, independently of their dependence on or care for future persons. The ability to ground sound, wide-ranging obligations of intergenerational justice is especially important given the urgent need to tackle climate change.

In the next chapter, I will draw my arguments together to formulate a provisional account of the scope of climate justice. I will then address three unresolved issues – the variety of intergenerational actions we engage in, whether the account can ground obligations to remote future generations, and the role of intergenerational cooperation within it – and use my conclusions to formulate a final revised version of the account.

Notes

1 Scheffler uses the term "afterlife," or "collective afterlife," in a non-religious sense to mean the existence of humanity on earth after we as individuals have died. In what follows I will use the expressions in the same way and interchangeably.

2 He does, however, think there may be exceptions. He thinks it is plausible, for example, that friendships and other personal relationship would still matter to people in an infertile world, as would playing games and – for different reasons – the afterlife itself (Scheffler 2013: 54–58).

3 He proposes three distinct ways in which what matters can depend on the afterlife (Scheffler 2013: 51). According to the "attitudinal dependency thesis," what matters to

us depends on our confidence in the existence of a collective afterlife. Only the attitudinal dependency thesis is given direct support by the afterlife conjecture. In addition, Scheffler proposes the "evaluative dependency thesis" ("the mattering *simpliciter* [of these things] depends on the actual existence of the afterlife") and the "justificatory dependency thesis" ("we are justified in attaching importance to things . . . only if there is an afterlife"); although he seems to support these additional theses, he concedes that they are only implied by the conjecture and may turn out to be false (Scheffler 2013: 52–53). Throughout this book, I follow Scheffler in assuming that the afterlife conjecture supports the attitudinal dependency thesis. My following arguments will rely on the afterlife mattering to us in this way.

4 For this distinction between different levels of moral theorising, see Stark 2010: 826–827. On Stark's account, general moral rules are triggered and applied through a process of deliberation. What I call the level of grounding she therefore identifies as the level of deliberation.

5 The question of what it means to treat others as mere means is at the centre of a larger and more complex debate and has given rise to a number of competing interpretations. O'Neill's is one prominent version but by no means uncontested. Engaging in this debate would be beyond the scope of this book, and for the sake of the argument I will rely on her interpretation as one that has been widely accepted. For competing interpretations, see Kerstein 2019.

6 See Gardiner 2017 for a similar argument.

7 I use climate change as the main example throughout, but the same reasoning could be applied to other ways in which we can exert influence over the future, for example, through long-term public investments or public debt.

8 This goal is itself insufficient to avoid the most threatening impacts of climate change (IPCC 2018; United Nations Climate Change 2021). For simplicity, and because it remains the official maximum goal Parties to the UNFCCC have committed to, I will use 2°C for the purpose of this argument; however, the same argument would work if substituting the lower target of 1.5°C mentioned in the Paris Agreement and the Glasgow Climate Pact.

9 This is a very simplified sketch of discussions surrounding the carbon budget, but it is enough for the purpose of this argument.

10 While I will not try to settle the issue here, I assume that there is a way to divide the budget fairly between various generations and the relevant actors within each generation.

11 See the mitigation scenarios assessed by the IPCC in IPCC 2014a, 2014b.

12 For the current purpose, I use "mattering to us in their own right" and "caring about them" interchangeably. Scheffler himself uses both expressions and comments that "[f]or the purposes of this discussion, what these concepts have in common is more important than the ways in which they differ" (Scheffler 2013: 17). I take both to mean that we are, to a certain extent, personally and/or emotionally invested in the future of humanity.

Reference list

Climate Analytics and New Climate Institute. 2021. 'The CAT Thermometer'. *Climate Action Tracker*. November. https://climateactiontracker.org/global/cat-thermometer/.

Driskell Tate, Ryan, Christine Shearer, and Andiswa Matikinca. 2021. 'Deep Trouble: Tracking Global Coal Mine Proposals'. *Global Energy Monitor and Oxpeckers*. https://globalenergymonitor.org/wp-content/uploads/2021/05/CoalMines_2021_r4.pdf.

Food and Agriculture Organization of the United Nations. 2020. *State of the World's Forests 2020: Forestry, Biodiversity and People*. Rome, Italy: FAO.

Frankfurt, Harry G. 2013. 'How the Afterlife Matters'. In *Death and the Afterlife, by Samuel Scheffler*, edited by Niko Kolodny. Oxford: Oxford University Press. https://doi.org/10.1093/acprof:oso/9780199982509.003.0006.

Gardiner, Stephen M. 2017. 'The Threat of Intergenerational Extortion: On the Temptation to Become the Climate Mafia, Masquerading as an Intergenerational Robin Hood'. *Canadian Journal of Philosophy* 47 (2–3): 368–394. https://doi.org/10.1080/00455091.2017.1302249.

IPCC. 2014a. *Climate Change 2014: Impacts, Adaptation, and Vulnerability. Working Group II Contribution to the Fifth Assessment Report of the Intergovernmental Panel on Climate Change*. New York: IPCC.

———. 2014b. *Climate Change 2014: Mitigation of Climate Change. Contribution of Working Group III to the Fifth Assessment Report of the Intergovernmental Panel on Climate Change*. New York: IPCC.

———. 2018. *IPCC Special Report Global Warming of 1.5°C*. Incheon, Republic of Korea: IPCC.

———. 2021. *Climate Change 2021: The Physical Science Basis. Contribution of Working Group I to the Sixth Assessment Report of the Intergivernmental Panel on Climate Change*. Cambridge: IPCC.

Kerstein, Samuel. 2009. 'Treating Others Merely as Means'. *Utilitas* 21 (2): 163–180. https://doi.org/10.1017/S0953820809003458.

———. 2019. 'Treating Persons as Means'. In *The Stanford Encyclopedia of Philosophy*, edited by Edward N. Zalta. Stanford: Metaphysics Research Lab, Stanford University. Summer.

Kolodny, Niko. 2013. 'That I Should Die and Others Live'. In *Death and the Afterlife, by Samuel Scheffler*, edited by Niko Kolodny. Oxford: Oxford University Press.

O'Neill, Onora. 1985. 'Between Consenting Adults'. *Philosophy & Public Affairs* 14 (3): 252–277.

Parfit, Derek. 2001. 'Rationality and Reasons'. In *Exploring Practical Philosophy: From Action to Values*, edited by Dan Egonsson and Ingmar Persson, 17–39. Burlington: Ashgate.

Scheffler, Samuel. 2013. *Death and the Afterlife*. Edited by Niko Kolodny. Oxford: Oxford University Press.

Stark, Cynthia A. 2010. 'Abstraction and Justification in Moral Theory'. *Hypatia* 25 (4): 825–833.

United Nations Climate Change. 2015. 'Paris Agreement to the United Nations Framework Convention on Climate Change'. https://unfccc.int/sites/default/files/english_paris_agreement.pdf.

———. 2021. 'Glasgow Climate Pact'. https://unfccc.int/sites/default/files/resource/cop26_auv_2f_cover_decision.pdf.

5 An account of the scope of climate justice

Taking stock

The question I aim to answer in this book is to what extent, and why, we have obligations of climate justice to future people. This is a question about the scope of climate justice, whereby "scope" defines "the range of persons who have claims upon and responsibilities to each other arising from considerations of justice" (Abizadeh 2007: 323). Determining the scope of climate justice is an important step towards constructing a complete, substantive account of climate justice. It is also necessary in its own right in order to fully address the intergenerational climate challenge.

In Chapters 2 to 4 I articulated three criteria that an account of the scope of climate justice must meet and proposed ways in which it can do so. I argued that an account of scope must provide measures to cope with uncertainty, be able to deal with complex moral problems like climate change, and include future people for their own sake. I also began to formulate an action-centred approach to scope that fits this framework and can explain that, and why, we have duties of climate justice to future people. In this chapter, I will draw together what I have discussed so far and propose a preliminary account of the scope of climate justice; I will then address some important issues this proposal does not fully resolve; finally, I will use this material to present my final, action-centred account of the scope of climate justice.

This chapter will tackle three issues that the preliminary account does not say enough about. The first has to do with the range of actions that are rightly described as intergenerational actions. I will argue that because our societies are intrinsically intergenerational, paying closer attention to our motivations for acting shows that intergenerational actions are more common than one might assume. One way to think about this is by applying Scheffler's afterlife conjecture to the account of scope. I discussed Scheffler's arguments at length in Chapter 4, arguing that our dependence on future people cannot serve as a justification for intergenerational justice. I will now revisit his arguments from a different point of view: I will argue that his conjecture points to an additional, wide-ranging category of possible intergenerational actions. Although, as argued in Chapter 4, this cannot be a justification for justice, these intergenerational actions can nevertheless ground duties to future people by triggering the account's coherence and normative requirements.

DOI: 10.4324/9781003258902-5

The second point has to do with the range of future generations who are included in the scope of our obligations on the proposed account. More specifically, I will discuss whether intergenerational actions can ground obligations to remote future generations in addition to near ones. In order to determine which generations are included by the account I will assess some of our intergenerational actions, showing that we often rely on presuppositions about both near and remote future generations. This suggests that the account can ground obligations to near as well as to more remote future people.

As a third and final point I will address the role of intergenerational cooperation within an account of scope. My aim is to determine the ways in which, if at all, engaging in strongly intergenerational actions, and hence in intergenerational cooperation, affects the scope of our obligations. I will approach this question by assessing the presuppositions on which different kinds of intergenerational actions rely, suggesting that it is the specificity of these presuppositions – that is, which specific capabilities are presupposed, beyond the mere capability for agency – that determines which future capabilities fall within the scope of our obligations. I will argue that intergenerational cooperation therefore indirectly broadens the scope of our obligations, as cooperative actions rely on more specific presuppositions about future capabilities and vulnerabilities than non-cooperative actions. I will conclude by revising the provisional account to include these specifications and proposing my final account of the scope of climate justice.

A preliminary account of scope

Previous chapters worked towards answering a key question about future generations and the scope of climate justice: do we owe duties of climate justice to future people, and if so, why? Drawing on what I argued so far I can formulate a provisional account of the scope of climate justice that provides an initial answer to this question.

Account (A1)

1   Our actions are based on certain presuppositions about others, including, often, about future people; these presuppositions determine whether and to whom we owe duties of justice and climate justice.
2   If our actions rely on presuppositions about future others, we must also recognise these future people as independent moral agents in our underlying principles of justice (*coherence requirement*) and extend the basic requirements of justice to them (*normative requirement*).
3   The scope of an agent's duties of justice therefore extends to all those present and future others whom their actions presuppose. If an agent is engaged in activities relevant to climate justice, the scope of their duties of climate justice also extends to all others whom their actions presuppose.

Activities relevant to climate justice are thereby defined as actions that directly affect the extent to which we are able to mitigate, adapt, or otherwise respond to anthropogenic climate change. They cover actions as well as omissions that are consciously performed and intended by their agents and include attitudes, policies, and practices. It is also important to recall the definition of presuppositions that is relevant in this context. When I say that an action is based on a certain presupposition about others, I mean that the presupposition explains certain features of the action in its current form and makes it conceptually coherent. Although they can be, presuppositions are not necessarily part of the agent's consciousness but rather a description of their action.[1]

For this account to be successful in fixing the scope of climate justice it must fit the framework developed in previous chapters and, in particular, meet what I argued are three necessary criteria: it must be able to cope with uncertainty, provide ways to deal with climate change as a complex issue, and include future people for their own sake. In the remainder of this section, I will look at how the proposed account meets these requirements.

Dealing with uncertainty

As a first criterion, I argued in Chapter 2 that *an account of scope must be able to accommodate uncertainties about the future effects of our climate changing and mitigating activities.*

Contrary to, for instance, its consequentialist counterpart, this account provides an action-guiding and normatively accurate way of dealing with uncertainty. That is, it provides the agent with useful practical guidance for the action under consideration and can accurately identify the morally relevant features of the action.

The proposed account of scope can successfully respond to the uncertainty of climate change for two reasons: (i) it takes each risk-imposition and its justification as the subject of moral evaluation and (ii) it takes risks to give rise to obligations of climate justice if they meet the following conditions: they are foreseeable by a reasonable agent at the time of acting and they expose others to risks to their autonomy-relevant conditions.

Defining a notion of foreseeable risk is a crucial step that allows us to apply the account to actions under uncertainty. The concept of foreseeable risk specifies that whenever a risk is foreseeable to the agent, the risk-imposing action is necessarily based on, among other things, weighing the preferences of the risk-imposing agents against those of the risk-exposed. This does not mean that it is necessary for the agent to consciously engage in this trade-off. Rather, it means that an action which involves a foreseeable risk is best described as involving the weighing of the preferences of the risk-imposing agents against those of the risk-exposed others. In other words, such an activity is based on presuppositions about the capabilities and vulnerabilities of the risk-exposed.

When the risk-exposed are future people, as in most actions related to climate justice, these risk-imposing actions are based on presuppositions about future people and are therefore intergenerational actions. By triggering the coherence and normative requirements they bring the future risk-exposed into the scope of the agent's obligations of climate justice.

At the same time, this account points us to one possible way of filling out the substance of our obligations to future people in a situation of uncertainty. By linking our obligations to an implicit trade-off between preferences and the presuppositions this trade-off rests on, it suggests that in situations of uncertainty, obligations of justice to future risk-exposed agents can be discharged by not giving undue weight to one's own preferences over the ability of others to enjoy the conditions of autonomy.

Dealing with complexity

As a second criterion, I argued that *an account of the scope of climate justice must be able to respond to climate change as a complex problem of justice.*

The complexity of climate change lies in a combination of specific problems – the dispersion of causes and effects, asymmetric vulnerabilities, as well as the severity, uncertainty, and scale of outcomes – which converge to make climate change an especially complex problem of justice. This limits the ability of many traditional moral theories to successfully determine whether and why we have duties of climate justice to future people.

By using a practical, action-centred methodology, the proposed account can meet the challenges of complexity and establish obligations of justice to future people that are based on agents' intergenerational actions. It can do this because its methodology only relies on our actions and their presuppositions to ground intergenerational obligations of justice. The necessary information about our actions remains accessible to us even in complex and uncertain situations like climate change.

This account responds to both Gardiner's and Jamieson's concerns about the inability of current moral theories to respond to climate change. By linking the traditional values of autonomy and respect for agency to the concept of intergenerational actions it highlights existing connections between our and future people's agency. These action-based bonds, in connection with the requirement to respect others' agency and autonomy, explain why we must include future people in the scope of our obligations of climate justice.

This responds to Gardiner's call for the development of our moral theories in relevant areas – including intergenerational justice – as a way to uncover the moral norms needed to respond to climate change. At the same time, it responds to Jamieson's worry about the inadequacy of existing concepts of responsibility for dealing with large-scale and diffuse problems like climate change, without however requiring us to develop and adopt new values or moral concepts. This account shows that the values we have do in fact permit us to access the necessary norms of climate justice if only we use other methods for applying them to the problem.

This approach places a lighter burden on present generations than an account which requires the adoption of new, alternative values. Moreover, merely extending the values of justice we already hold removes the degree of abstraction inherent in the development of new values and concepts. This contributes to streamlining the resulting theory of climate justice and simplifying its fit with existing commitments in other domains of justice, making it more action-guiding in a time of crisis.

Including future people for their own sake

As a third and final criterion, I argued that *an account of the scope of climate justice must include each person for their own sake and as an end in themselves*. This criterion ensures that future agents are never treated as mere means to our ends and that principles of intergenerational climate justice promote just future worlds. I argued that, in order to achieve this, the only acceptable justification for justice is an agent's standing as an end in themselves and never their role in furthering someone else's ends.

The requirement to include each person for their own sake is at the core of the proposed account. It is embodied by the coherence and normative requirements, which spell out the key normative elements of the account. According to the coherence requirement, whenever our actions rely on presuppositions about others our principles of justice must recognise these others as independent moral agents. The normative requirement then ascribes each agent an equal entitlement to the conditions of autonomy and to being respected as an agent. This requirement is what provides the justification for the duties of justice the account gives rise to.

Prior to applying the coherence and normative requirements, the account uses our intergenerational actions to identify the range of others on whom our actions rely and whom we must therefore recognise as moral agents in acting. Our actions thus ground our obligations of justice to those others but are not what justifies them. Our reliance on others as part of our intergenerational actions requires us, through the coherence requirement, to recognise these others as agents, yet our obligations to them are then justified by the normative requirement to value these others as independent moral agents who are equally entitled to the conditions of autonomy.

On the proposed account, future people are thus never included in the scope of justice as mere means to our intergenerational actions but as moral agents on whom our actions rely and who are, as independent agents, entitled to the basic requirements of justice.

Scheffler's afterlife conjecture and intergenerational actions

The afterlife conjecture

The preliminary account of scope fits the developed framework and meets the necessary criteria for a successful account of scope. However, it is silent, or says too little, about three important issues. Let me begin by tackling the first of these, which is the range of present actions that count as intergenerational actions.

In Chapter 3, I suggested that there are two broad types of intergenerational actions: stronger or cooperative intergenerational actions, which require future people to contribute in order to be completed successfully; and weaker intergenerational actions, which do not rely on future people for completion but are nevertheless based on presuppositions about them. As possible examples I mentioned the practice of public debt and long-term climate plans on the one hand and future-oriented legal documents like the Paris Agreement on the other hand. I also suggested that our societies are intrinsically intergenerational and that this intergenerationality provides the necessary background against which many of our intergenerational actions are made possible.

On Samuel Scheffler's interpretation of what an intergenerational society means in practice, this intergenerationality implies that a much wider range of actions than those just mentioned are in fact intergenerational. I already discussed Scheffler's afterlife conjecture in the previous chapter. Scheffler argues that past, present, and future generations are connected as parts of the ongoing project of humanity and that unless we can think of ourselves as parts of this continuum, many of even the most trivial objects and pursuits we value as part of our daily lives will cease to be important to us. The reason behind this is that we need the frame of reference provided by humanity as an ongoing enterprise in order to develop a sense of what is valuable and important. If, instead, we were to be confronted with the imminent extinction of humanity after our death – the loss of the "collective afterlife" – this would compromise our ability to value many of the activities that had previously seemed worthwhile to us, even mundane and recreational ones (Scheffler 2013). According to Scheffler, his conjecture shows that our ability to live valuable lives here and now largely depends on our belief that humanity will continue to exist for "a healthy and indefinitely long period after our own deaths" (Scheffler 2013: 63).

I argued in Chapter 4 that the dependence on future people which Scheffler describes cannot serve as a justification for including future people in the scope of justice, as doing so would risk using future people as mere means and sanction unjust worlds. Instead, we ought to include future and present persons for their own sake and justify their inclusion in the scope of justice with reasons that apply equally to all. The proposed account meets this criterion by justifying our duties to future people with every agent's equal entitlement to respect and the conditions of autonomy.

Intergenerational actions, however, are also crucial to the proposed account. They provide the legitimate grounds through which to identify which future agents the normative requirement applies to. Thus, if Scheffler is right in arguing that so many of our activities are dependent on the existence of future people, his suggestion can complement and have wide-ranging implications for an account of scope based on intergenerational actions. In turn, my proposed account can supplement Scheffler's conjecture by showing one way in which our dependence on future people grounds duties of intergenerational justice, while still including future people for their own sake. Let's unpack why this is the case.

Three categories of intergenerational actions

According to Scheffler's conjecture, many seemingly present-oriented actions are in fact intergenerational. Supposedly, the value and importance of these actions to us is part of the reason why we engage in them, and they can only be valuable and important to us under the precondition that humanity will continue to exist after our deaths. Scheffler's speculation seems plausible, and I have so far assumed it to be correct. If indeed it is, it seems that a much wider range of actions than I have so far identified rely on presuppositions about future people, thus triggering the coherence and normative requirements to bring future people into the scope of the agents' duties of justice.

Scheffler mentions three categories of actions that he believes would become less valuable to us without the prospect of a collective afterlife. The first are activities that would already be described as intergenerational actions on the proposed account of scope: they include projects oriented towards achieving a future goal such as technical or medical research; creative projects that depend, at least in part, on their reception by a future audience; activities that aim to sustain certain values and practices over time, such as participating in traditions and rituals; and activities aimed at promoting the survival of particular cultural or communal groups over time (Scheffler 2013: 42). Clearly, these activities seem to presuppose the existence of future people that can enjoy or make use of their results. Some of these, for example, long-term research projects carried out over multiple generations, may even require the contribution of future people to be successful.

What is more interesting is Scheffler's suggestion that, on the one hand, other less obviously intergenerational activities would also come to matter to us less and that, on the other hand, our very ability to value things would itself be compromised. He believes, first, that

the imminent disappearance of human life would exert a generally depressive effect on people's motivations and on their confidence in the value of their activities – that it would reduce their capacity for enthusiasm and for wholehearted and joyful activity across a very wide front. (Scheffler 2013: 43)

This loss of enthusiasm would affect things like enjoying nature or music and the pursuit of intellectual activities. The reason why we would struggle to still see these actions as important to us is that their value to us depends on our ability to confidently place them within an ongoing human history, and we are only able to do so if we can be confident that human beings will continue to exist after we die and for some time thereafter. If Scheffler's afterlife conjecture is correct, it therefore looks like a much wider range of very common, seemingly present-oriented activities are in fact based on presuppositions about future people.

As a third category, Scheffler argues that even our understanding of what it means to value something, of things and activities being in some way important to us, can only retain its meaning under the precondition that we ourselves are part of an ongoing human history. The practice of valuing

occurs within the implicit framework of a set of assumptions that includes, at the most basic level, the assumption that human life itself matters, and that it is an ongoing phenomenon with a history that transcends the history of any individual. (Scheffler 2013: 59)

Thus, according to Scheffler's conjecture, the act of valuing as we know it would be conceptually incoherent unless we presupposed the existence of future people for a sufficiently long period of time after our own deaths.

It is worth briefly pointing out that the importance Scheffler ascribes to the collective afterlife as a critical component of our shared history is specific to, or at least modelled after, the relationship that a linear conception of time presupposes with the future. That is, the desire or even need for a collective afterlife is likely to be central to societies whose cultural conception of time is linear and future-oriented and less prominent in cultures with a cyclical notion of time for whom the past may play an equal if not more important role. Although these considerations in no way invalidate Scheffler's conjecture or arguments that are based on it – many cultures and global and international political organisations do follow a linear conception of time – it is important to keep in mind that these arguments may not apply equally to societies across the globe.

The afterlife conjecture and the scope of justice

Taking a closer look at Scheffler's afterlife conjecture reveals that two additional types of activities which had not been identified by the account so far are in fact, and quite plausibly, intergenerational actions that rely on presuppositions about future people. This has important implications for the reach of our account of the scope of climate justice.

First, and importantly, most agents regularly engage in these activities. Most individual agents regularly engage in mundane activities like listening to, playing, or enjoying music; walking in a park; enjoying beautiful scenery; or learning a new language or skill. All individual agents will also have valued things or pursuits in their lifetimes and have perceived various objects, activities, or states as being important to them.

We can reach a similar conclusion for the collective agents relevant to climate justice that are the focus of this book. On the understanding of collective agents that I am working with, we can make sense of the idea that something can be valuable or important to a collective agent too. Recall that, on this view of collective responsibility, organised collectives that meet specified criteria can form a collective will, exercise agency, assume moral responsibilities, and be attributed collective intentions that can differ from the individual intentions of their members.[2] The coherence and normative requirements thus apply unaltered to these organised collective agents, who, like individual agents, acquire an obligation to respect others' agency whenever their actions rely on it.

Insofar as what is important to human agents in general depends on their ability to place themselves within an ongoing human history, the same applies to such organised collective agents. Assuming, as Scheffler's conjecture does, that agents have an interest in their projects and activities persisting after they are gone, this plausibly applies to the collective agents in question as much as to individual agents. In turn, collective and individual agents are likely to have a strong interest in a collective afterlife to ensure they and their projects are part of an ongoing human history.

A range of common activities by collective agents are therefore vulnerable to the afterlife conjecture. For a collective agent as I have described it, it can for instance make sense to say that something is valuable or important to them if it is in line with their national, communal, or organisational culture or with their ethos, policies, or strategy. Similarly, we can imagine daily, mundane activities of collective agents which would be affected by the afterlife conjecture to be things like entering or exiting political unions or partnerships; passing new legislation; developing new business or communication strategies; or launching and advertising new products. These are only somewhat general speculations, but they highlight sufficient relevant similarities to suggest that, if the afterlife conjecture is plausible for individual agents, there are reasons to think it is plausible for organised collective agents as well.

If this is so, we can expect that activities like those just mentioned, as well as the more general idea of something being important to collective agents, would be affected by the knowledge that humanity will soon be extinct. Most collective agents, including those relevant to climate justice, will regularly engage in such actions. The European Union, for example, may negotiate the terms of entry for a new member state, or, to rely on a timely example, has not long ago negotiated the terms of exit for the UK; on the other hand, we can say that a local government values the erection of a statue to commemorate its founder or that a political party committed to social and economic justice values the introduction of mandatory gender and minority quotas in private companies.

Second, although the presuppositions about future agents on which the afterlife conjecture relies are quite minimal, they are enough to trigger the coherence and normative requirements of the account. Neither our daily activities nor our ability to value things must necessarily rely on particularly detailed presuppositions about the specific capabilities and vulnerabilities of future people. Yet, on the proposed account of scope, any presuppositions about future people as agents, regardless of how minimal they are, are enough to bring them within the scope of the agent's obligations of justice. Regardless of how little our actions presuppose about future people and how little we know about their future preferences, interests, and capabilities, it is plausible to say that, as human agents, they too will be entitled to the basic conditions of autonomy called for by the account's normative requirement.

Hence, if the proposed approach can work for other intergenerational actions, and provided Scheffler's afterlife conjecture is plausible, the account should apply to these additional intergenerational actions too. In turn, this means that agents who engage in these activities – which, as I argued, are nearly all agents – acquire intergenerational obligations of justice.

Which future generations?

This brings up a second important issue that requires clarification: how temporally far-reaching are the obligations that agents who engage in intergenerational actions acquire under the account? And, specifically, can intergenerational actions also serve to bring the tougher case of remote future generations into the scope of (climate) justice?

The preliminary account provides a partial answer to these questions. It suggests that the temporal reach of agent's obligations depends on how far-reaching their actions' underlying presuppositions about future people are and hence on the specific intergenerational action the agent engages in. For a more conclusive answer, we therefore need to take a closer look at possible intergenerational actions.

Near future generations

Let me begin with actions that rely on presuppositions about near future generations. As near future generations I understand future people that are closer to us in time, that is, people who are going to live within about a century from the present. Most of the intergenerational actions discussed in the previous chapter are based on presuppositions about future people that fall within this approximate range.

The Paris Agreement, for instance, often relies on presuppositions until the end of this century. It states in Article 4, for example, that in order to achieve its long-term temperature goal Parties to the Agreement should aim to reach net zero emissions by the second half of this century (United Nations Climate Change 2015: Art. 4.1). The Agreement also requires Parties to communicate their NDCs every 5 years (United Nations Climate Change 2015: Art. 4.9) and to undertake a global stocktake on progress towards the goals of the Agreement every 5 years starting in 2023 (United Nations Climate Change 2015: Art. 14). While the latter two requirements do not include an upper time bound, given the overarching aim to reach net zero emissions within this century it also makes sense to interpret these efforts as focused mainly on the current century.

Similarly, Germany's Climate Action Plan 2050 and its 2030 climate action package – which were also discussed in Chapter 3 – set out measures and targets for, and thus rely on presuppositions about, the next 10 to 30 years (Bundesministerium für Umwelt, Naturschutz und nukleare Sicherheit 2016; Deutscher Bundestag 2019). A similar timespan plausibly applies to intergenerational actions such as large-scale building and infrastructure projects or investment and economic policy decisions. Overall, there are many intergenerational actions that ground obligations to near future generations.

Remote future generations

At the same time, we engage in activities that are based on much more long-term presuppositions about the future. I have in mind, on the one hand, Scheffler's afterlife conjecture and its related intergenerational actions and on the other hand the storage of toxic nuclear waste. These actions also bring remote future generations into the scope of justice.

If we accept Scheffler's afterlife conjecture, many of our most common activities, including the act of valuing itself, are only meaningful if agents can view themselves as parts of a common human history. This means that these actions must be based on the presupposition that there are going to be near and remote future generations of humans.

Scheffler does not specify how long we should want humanity to survive for. He only speaks of a "healthy and indefinitely long period after our own deaths" (Scheffler 2013: 63). We may, however, be able to narrow down this time period a little further. It is estimated that the earth will start to become inhabitable in around a billion years (Leconte et al. 2013), but it is likely that humanity will become extinct much earlier. To provide some relevant context, mammal species have an average lifespan of about 1 million years, though some have existed for up to 10 million years (Evolution Library 2001). Estimates for the species lifespan of Homo Sapiens vary considerably: scientists have estimated the probability of human extinction due to natural causes at less than one in 14,000 for any given year, but predictions are complicated by the fact that anthropogenic causes of extinction are thought to be more likely than natural ones (Snyder-Beattie et al. 2019).

For the purposes of the afterlife conjecture, however, we may not even need to be confident that humanity will continue to exist for millions of years. Perhaps, though this is only my speculation, we can plausibly say that, in order to satisfy our need for a collective afterlife which frames and gives meaning to our actions, we need to be confident that humanity will continue to exist and thrive for at least another thousand years. This time frame seems long enough to allow for a meaningful continuation of our societies and traditions yet short enough to ensure societies and people within this time frame are sufficiently similar to us, in at least some of the most fundamental aspects of life.

Our notion of future humans becomes increasingly abstract the further we move ahead in time, while, at the same time, it is also increasingly likely that more remote future societies will be significantly different from the ones we are accustomed to. Part of what it means to value the continuation of a practice or tradition is presumably to value its continued relevance in the future. In turn, the reasons we have for valuing certain objects or pursuits include their relevance to our lives so that at least part of the value of their continuation would seem to be lost once they become irrelevant to future people's lives.

It is reasonable to expect that such changes in relevance will be minor at first but that, as time passes, people and societies will have evolved to such an extent as to value different practices and traditions altogether. While this does not mean that societies existing more than a millennium from now will necessarily be so different as to be unable to continue our traditions and projects or to value our art and culture, it does suggest that we may not need to presuppose the appreciation by such remote societies in order to be confident that our societies, practices, and pursuits are not unduly cut short. This timeline need not necessarily apply to all cases in practice – it may be shorter or longer, depending on the values that are at stake – but, for the purpose of the argument, it seems to be a reasonable time frame to work with.

A second set of actions which rely on presuppositions about the existence of remote future people is our efforts to safely store toxic nuclear waste. This example may help put our notion of the very remote future into perspective and substantiate an appropriate timespan we can work with. Some of the radioactive isotopes produced as the by-products of nuclear fission continue to emit dangerous radiation for thousands of years before decaying: plutonium-239, for example, has a half-life of 24,000 years (United States Nuclear Regulatory Commission 2019). This and other nuclear waste products can be stored in special repositories that are designed to safely store radioactive waste for especially long periods of time.

The Waste Isolation Pilot Plant in New Mexico, USA, for example, is designed to store nuclear waste for at least up to 10,000 years (Howard et al. 2000). Efforts to safely store toxic nuclear waste have also given rise to an entire field of research – called nuclear semiotics – devoted to developing effective ways of communicating the dangers of nuclear waste depositories to very remote future generations, who are likely not to share our languages, writing, and understanding of visual signs (Conca 2015; Stothard 2016).

Actions to safely store nuclear waste in the long-term and inform future generations of its dangers seem to be based on the presupposition that there are likely going to be people in the very remote future who will be vulnerable to nuclear radiation and able to comprehend some form of visual or textual message about its dangers. These actions, as well as activities affected by the afterlife conjecture, are some possible examples of actions that bring remote future generations – ranging from about 100 to 10,000 years from now – into the scope of agents' obligations of justice. Agents who engage in these actions will acquire an obligation to respect at least the basic conditions of autonomy of the relevant remote future people. For our account of scope this means that, depending on the actions each agent engages in, their obligations can be temporally very far-reaching and include both near and remote future generations.

Intergenerational cooperation and the scope of justice

Let me now turn to the third and final issue, namely the role that intergenerational cooperation plays within our account of scope. More specifically, I am interested in the extent to which engaging in intergenerational cooperative actions may affect the scope of an agent's obligations. This clarification is particularly important given the range of different, and variously strong, intergenerational actions that are relevant to the account.

I argued in the previous chapter that one particular kind of intergenerational actions – strongly intergenerational actions – which rely on the contribution of future people to be completed successfully also connect the participating generations in a type of intergenerational cooperation. Examples of these actions are the practice of public debt, construction projects spanning longer than one generation and the implementation of long-term climate action plans. The element of intergenerational cooperation is entailed in the activities themselves: though at different times, multiple generations are required to contribute to the same project in order for it to be completed successfully.

As a final clarification before I formulate the final account of scope, this section will investigate the role that intergenerational cooperation plays within the account and the effect it has on the scope of agents' obligations, if any, compared to non-cooperative intergenerational actions. One way to approach these questions is to think about the kinds of presuppositions that weaker intergenerational actions on the one hand and stronger, cooperative intergenerational actions on the other hand are based on. Doing so suggests that, first, the specificity of the relevant presuppositions affects the resulting duties and, that, second, the distinction between weak and strong intergenerational actions only represents two extremes of what is in fact a much broader spectrum of intergenerationality.

As outlined in Chapter 3, the various capabilities our actions may presuppose in others – the capability for agency, as well as more specific capabilities I will discuss in greater detail later – are relevant to this account as binary properties rather than as a matter of degree. That is, the coherence and normative requirements are triggered whenever an action presupposes a certain capability, regardless of the level at which that capability is presupposed. When I speak of the specificity, strength, or preciseness of presuppositions I am therefore speaking of the types of capabilities that are presupposed, not the degree to which they are presupposed.

Weaker intergenerational actions

What I have described as weaker intergenerational actions are those actions that, though not reliant on future generations for completion, are nevertheless based on the presupposition that there are going to be future people with certain capabilities and vulnerabilities. One of the intergenerational actions introduced in this chapter – the storage of nuclear waste – is a good example of this kind of activity.

To be completed successfully, the act of safely storing radioactive material does not require future people to conduct themselves in any specific way. Hence, to be conceptually coherent, this action only relies on relatively minimal presuppositions about the future. Its underlying presuppositions – the presuppositions which make it, in its current form, a coherent action – only concern the existence of future people with basic physical and mental human features, including, most obviously, the vulnerability to nuclear radiation.

According to the coherence requirement of the account, the presuppositions underlying agents' actions must also inform the principles of justice that guide these agents' activities. The relevant presuppositions on which such weakly intergenerational actions rely are general presuppositions about future people as agents: they do not rely on any presuppositions about future agents' specific capabilities and vulnerabilities. These minimal presuppositions are enough to trigger the normative requirement. According to the normative requirement, because these actions rely on presuppositions about future generation, agents who engage in them acquire an obligation to respect the basic entitlements of justice of these future people. In other words, they acquire an obligation to respect their conditions of autonomy. As outlined in Chapter 2, these include goods such as life, shelter, and nutrition and are so fundamental to human lives that we may assume them to be relevant to remote future generations, too.

Because weaker intergenerational actions of this kind do not rely on any more precise presuppositions about future capabilities and vulnerabilities, on the proposed account these actions do not give rise to obligations with regard to future people's specific capabilities and vulnerabilities. In other words, when dealing with weaker intergenerational actions, only the basic requirements to respect future agents and their autonomy conditions fall within the scope of agents' obligations of justice.

Stronger intergenerational actions

Stronger or cooperative intergenerational actions, on the other hand, rely on much more specific presuppositions about the capabilities and vulnerabilities of future people. I described these actions as based on particularly strong presuppositions about future people. By this I mean that, because they are based on the expectation that future people will at least be able to continue the activity in question, these actions must rely on the presupposition that future people will have the specific capabilities necessary to do so.

Actions like implementing long-term climate change mitigation and adaptation plans, or executing large, long-term construction projects, are all based on the presupposition that future people will be able to contribute to the activity in question. In turn, agents that initiate or participate in these actions do so based on the presupposition that future people will have the necessary capabilities to advance a construction project or adhere to climate change plans and that they will be at least likely to use their capabilities to contribute to these intergenerational actions.

The specificity of these actions' underlying presuppositions differentiates strongly intergenerational actions from weaker ones. What makes strongly intergenerational actions conceptually coherent is not just the presupposition that there are going to be future people with the basic capacity for agency but the additional presupposition that there are going to be future people with the necessary capabilities to engage in the activity in question.

Recall that the coherence and normative requirements hold that, because some actions presuppose future people, agents who engage in those activities acquire an obligation to honour the basic autonomy conditions these future people are owed as agents. What if, however, an action presupposes more than just the existence of future people and instead relies on the existence of future agents with certain, more specific capabilities?

The normative requirement builds on what the action presupposes to accord future persons their entitlements of justice. It therefore ought to be the case that, in cases in which the presuppositions are more specific – as in cooperative intergenerational actions – agents acquire additional obligations to respect the more specific capabilities on which their actions rely. In other words, given the coherence and normative requirements, it follows that if an action is conceptually reliant on presuppositions about future people as agents with specific capabilities, then respecting these more specific capabilities, in addition to the basic conditions of autonomy, will also fall within the scope of the present agent's obligations. Intergenerational cooperation can therefore be said to indirectly affect our obligations by broadening their scope due to the specificity of the presuppositions on which cooperative actions necessarily rely.

A nuanced approach to intergenerational actions

If the account is in this way sensitive to the strength of the presuppositions on which stronger and weaker intergenerational actions rely, it must also be able to account for the various nuances of intergenerational actions that lie between very weak intergenerational actions on the one hand and cooperative intergenerational actions on the other. In other words, the sensitivity just outlined suggests that the entire breadth of agents' obligations needs to be calibrated depending on the strength of their actions' presuppositions about future people.

What this means in practice can be seen by looking, for example, at the Paris Agreement. Earlier in the book I discussed the Agreement as an important instance of a weaker, non-cooperative intergenerational action. The aspect of the Agreement I focused on for the sake of the argument was not its implementation – for which it would need the contribution of future people, just like the national climate action plans I discussed – but its authoring, signing, and ratification by member states. These actions do not rely on future people for completion, yet they do rely on presuppositions about the capabilities and vulnerabilities of future people.

These are all activities that member states were able to successfully complete without the contribution of future generations. They are therefore unlike the stronger, cooperative actions discussed earlier. Nevertheless, because of the future-oriented structure and clear, future-oriented temperature and policy goals of the Agreement, these actions still rely on quite specific presuppositions about future people. They presuppose future people who will be vulnerable to the effects of climate change and who will be capable of taking the necessary mitigation measures to limit the rise in temperature.

These presuppositions make the Paris Agreement a particularly interesting intergenerational action. Although its authoring and ratification are not cooperative actions, the mitigation capabilities they presuppose in future people are significantly more specific than the capability to understand a basic textual or visual danger warning presupposed by nuclear semiotics or the vulnerability to radiation presupposed by the storage of nuclear waste, both of which can be classed as weaker intergenerational actions.

Acts of valuing, too, are a good example of a set of actions that lie somewhere between very weak and very strong intergenerational actions on the spectrum of intergenerationality. According to Scheffler, the meaningfulness to us of acts of valuing relies, at least in part, on our confidence in the existence of a collective afterlife. This claim is based on Scheffler's view of valuing as diachronic, that is, as creating "commitments that endure through the flux of daily experience" and as involving the wish to preserve the valued object over time (Scheffler 2013: 61).

This understanding of valuing, I believe, should not be read as relying on future people to preserve our valued objects and pursuits indefinitely. Valuing something may involve a desire for the original valued thing to be remembered, but, at the same time, it also seems to involve the recognition that values must change to remain relevant to future values. Seana Shiffrin makes a similar suggestion in her comments on Scheffler's lecture, where she argues that "it seems to matter greatly that the general practices of acknowledging value and acting on reasons continue, even if many of the particular, instantiated objects of that abstract activity alter and evolve" (Shiffrin 2013: 151–152).

A diachronic concept of valuing is vulnerable to the afterlife conjecture because it presupposes future people with a capability for valuing goods, pursuits, practices, and traditions, as well as for understanding and perhaps carrying on what we valued. While the act of valuing does not require future people to contribute to it to be successful – we can successfully value something even if future people do not – and although it presupposes less detailed capabilities than the Paris Agreement, it nevertheless presupposes more about future people than the basic human capabilities which are at the basis of very weak intergenerational actions. On the gradient of intergenerationality, it seems to fall somewhere between the storage of nuclear waste and the Paris Agreement.

All of this suggests that the distinction between cooperative and non-cooperative intergenerational actions is not exhaustive. I argued that cooperative intergenerational actions give rise to especially wide-ranging obligations, as their presuppositions about future people are, by necessity, particularly specific. However, the examples of the Paris Agreement and the practice of valuing suggest that the distinction between weak and strong intergenerational actions is not clear-cut but rather a matter of degree. All nuances of intergenerationality, which can vary from action to action, are equally relevant for determining the scope of agents' obligations. In each instance, it is the specificity of an action's underlying presuppositions which determines the scope of the obligations it grounds.

Two conclusions follow once we acknowledge that the account calls for a more nuanced approach to intergenerational actions. On the one hand, it is the case that every intergenerational action imposes an obligation on the agent to respect the basic conditions of autonomy of the future people it presupposes. On the other hand, the presuppositions underlying each action determine whether the scope of the agent's obligations extends beyond an obligation to respect future people's basic conditions of autonomy. The specificity of the agent's obligations depends on and tracks the specificity of their action's presuppositions and not just whether the action is cooperative or not.

An account of the scope of climate justice

The account, revisited

I have now clarified what I suggested were three important issues that the preliminary account had said nothing, or too little, about: I argued that many daily, mundane actions are in fact intergenerational; that our obligations can extend to remote and near future generations; and that the breadth of our obligations depends on the specificity of our actions' underlying presuppositions. With these clarifications in mind, I can now formulate the final version of my account of the scope of climate justice.

Account (A2)

1   Our actions are based on certain presuppositions about others, including, often, about future people; these presuppositions determine whether and to whom we owe duties of justice and climate justice.
2   If our actions rely on presuppositions about future others, we must also recognise these future people as independent moral agents in our underlying

principles of justice (*coherence requirement*) and extend the basic requirements of justice to them (*normative requirement*).

3   The scope of an agent's duties of justice therefore extends to all those present
    and future others whom their actions presuppose. If an agent is engaged in
    activities relevant to climate justice, the scope of their duties of climate jus-
    tice also extends to all others whom their actions presuppose.
4   The scope of an agent's obligations can extend to near and remote future
    generations, depending on the presuppositions their actions are based on.
5   The kinds of obligations that fall into the scope of an agent's duties – whether
    they include only a duty to respect future people's basic conditions of auton-
    omy or extend to more specific capabilities – depend on the specificity of the
    presuppositions on which the agent's actions are based.

The core elements of the account (1 to 3) have not changed. In addition to its first
formulation, however, the preceding discussion has allowed us to specify three
things: the temporal reach of the account (4); the extent to which the scope of
agents' obligations is determined by the strength of their presuppositions (5); and
the range of actions which, as intergenerational actions, can ground obligations
to future people.

While I had already argued that many of the specific actions engaged in by
collective agents relevant to climate justice are intergenerational – for example,
the Paris Agreement and many climate change mitigation and adaptation plans –
Scheffler's afterlife conjecture suggests that a large number of very common,
often daily, actions are in fact intergenerational too.

Because there is such a broad range of intergenerational actions, this account
of scope likely grounds obligations of climate justice for all – and at least for
the vast majority of – collective agents relevant to climate justice. By looking at
the actions that agents already engage in, it successfully explains to what extent,
and especially why, we have obligations of climate justice to future generations,
allowing us to successfully set the scope of climate justice.

The non-identity problem and non-existence challenge

As a final point, I want to suggest that the proposed account may also allow us
to circumvent two of the main challenges that have been raised against intergenerational duties of climate justice, namely the non-identity problem and the non-
existence challenge.

Let me start by addressing the non-existence challenge. This objection poses
the somewhat simpler challenge: it states that future people cannot, because of
their non-existence, be bearers of rights and interests in the present and that, therefore, theories of justice which require its subjects to have interests or rights at the
time at which the respective obligation is to be discharged cannot include future
people within their scope. I say that this is a simpler challenge because, even
if we concede the premise that future people cannot possess interests of rights
right now, it can be circumvented by any theory of justice which does not ground

obligations on the present interests of its subjects. Because the account of inter-generational obligations proposed in this book does not require its future subjects to have interests or rights at this moment in time, it allows us to overcome objections based on non-existence.

Similar approaches to solving the non-existence challenge are common in the existing literature on intergenerational justice. Feinberg's is one of the earliest contributions to this discussion. He argues the following: while the identity of non-existing, future people is undetermined and therefore unknown to us, we do know that once future generations come into existence their members will possess certain important interests; these interests, in turn, will ground a set of rights that we can infringe with our present, future-affecting actions (Feinberg 1974: 65). Feinberg's view was later echoed by Elliot (1989) and Meyer (2016). According to both theorists, what matters is that we can plausibly assume that future people are going to have interests once they come into existence and that these interests are the type of interests that give rise to obligations of justice. This approach also underpins Caney's recipient-based account of climate justice. Caney holds that although future people do not have human rights right now, they will in the future, and what matters is that these rights are liable to be violated by the actions we take now (Caney 2017). It seems to be at the basis of O'Neill's own thinking about intergenerational justice, too. She argues that while we cannot individuate members of future generations, we are able to specify future people as agents with finite capabilities and vulnerabilities, whom we can connect to through future-affecting actions. This is enough for us to make presuppositions about them, as our future-affecting actions show (O'Neill 1996: 115).

Like O'Neill's, my proposed account grounds our intergenerational duties on the presuppositions we make about future persons. The range of intergenerational actions discussed in earlier chapters proves O'Neill's claim that we only need to specify, not individuate, future people in order to base our actions on presuppositions about them. When we take on public debt, start a long-term construction project, or adopt a new climate plan, we only need to presuppose agents whom we can specify as members of future generations and who, as future people and independently of their identity, will possess certain interests and entitlements. Our obligations are justified by the entitlements to respect and autonomy these agents will possess once they come into existence. These entitlements and the underlying interests do not have to exist at the same time as our presuppositions and, in turn, our obligations. Instead, the account requires us to respect them as fundamental interests held by all agents, including, importantly, future people in their future lifetimes. Although neither the subjects of our obligations nor their interests or entitlements exist at the time of acting, as we engage in intergenerational actions we can shape our actions and their underlying presuppositions in ways that are respectful of future agents and which contribute to creating an environment that promotes their future conditions of autonomy.

The fact that the future subjects of justice do not have to exist at the time of our presuppositions renders the proposed account of scope more dynamic and ensures its ability to adapt more flexibly to society's changing need for justice.

Our societies are likely to undergo significant changes in the future, including the very near future. We may, for example, soon be able to envisage a growing participation of artificial intelligence and consequently increased interactions between human agents and artificial intelligence. On the proposed account, these or similar changes to our actions and their underlying presuppositions can pave the way for adapting the scope of our obligations, too, to mirror important social developments.

The differentiation between individuating and specifying future agents is also, in part, what allows the proposed account to circumvent the non-identity problem. The non-identity problem is the objection which poses the more serious challenge to intergenerational justice. As I laid out in Chapter 1, some theorists believe this puzzle eliminates the possibility of intergenerational climate justice altogether. Remember what the non-identity problem says: it holds that an action which does not make a person worse off, and which is a necessary condition for that person's existence, cannot be wrong because of how it affects that person, her interests, or her rights (Parfit 1984).

Non-identity considerations also bear on the case of climate change: those future people who will be exposed to the harmful consequences of climate change would not exist if we acted so as to avoid climate harms – if we committed to a less emitting lifestyle, adapted our lives accordingly, and hence conceived children at different times or with different partners. Those particular future people will therefore not be harmed, in the sense of being made worse off, by our climate changing activities. To the contrary, our allegedly harmful activities are in fact a necessary condition for their existence.

Theorists that are concerned with non-identity considerations see them as an obstacle to intergenerational duties of justice as follows: they argue that an action cannot constitute an injustice towards a person if it does not make her worse off, but if instead it benefits her by causing her to exist (see Broome 1992, 2012). By pointing to this discrepancy, the non-identity problem can affect the way in which we think about our relation to future generations. It has become the source of a large debate, leading a number of theorists to propose possible solutions to it;[3] throughout this book, I have assumed that among these there are successful ways in which the problem can be avoided.

Because the non-identity problem has been the source of such a significant philosophical disagreement, I am not hoping to provide a conclusive answer to it in this short space, nor one that will satisfy all its proponents. Nevertheless, because it is a critical issue to at least part of the philosophical community, a book on intergenerational climate justice must address it. I also believe my proposed approach offers a way to circumvent the non-identity problem and thus an additional possible response to theorists who see it as a hurdle to establishing intergenerational duties of climate justice.

The way in which the proposed approach can dodge non-identity considerations is by avoiding the use of harm as the ground for duties of justice, relying instead on an assessment of actions and their presuppositions about specified, but not individuated future people. The way I have suggested we think about intergenerational justice is not as a way to address the harms done to future people but as a way to honour and respect the fundamental needs – or autonomy conditions – of all persons, including future people.

While the non-identity problem rests on linking justice to harm, I argued that we should instead think of justice as requiring us to respect future people and their agency. Future people will be entitled to respect for their agency and their conditions of autonomy not as specific individuals but by virtue of their existence as persons. In other words, their entitlements and our corresponding duties are linked to their existence as persons but are at the same time identity-independent. If respecting those general entitlements is what justice demands of us, there is no need to show we can harm non-existing future individuals. Instead, we ought to consider the conditions in which future agents may exist and think of our actions and presuppositions as tools for constructing a just environment for them. One avenue for doing so is to act in ways that respect the agency and autonomy conditions of future people.

Whether we are duly respecting future agents and their entitlements to the conditions of autonomy can be read out of our actions and the presuppositions on which they are based. If we conceive of our obligations of justice in this way – as duties to respect future people and their autonomy conditions – acting justly towards future people does not require the subjects of justice to be individuated. When agents engage in the relevant intergenerational actions, they are able to act on presuppositions about unindividuated future people, specified only as future agents who are likely to be affected by, or expected to contribute to, their actions; equally, agents do not need to individuate future persons in order to act in ways that respect their future conditions of autonomy and promote a just future environment (see O'Neill 1996: 115–118).

This approach falls into a category of existing solutions to the non-identity problem that propose to solve the puzzle by distinguishing between what we may call types and tokens of a particular category – in this case, of future people. Types describe what I, following O'Neill, have called specified but not individuated future people. These future people, simply by virtue of being future people, are going to possess certain properties, interests, and consequently entitlements. Tokens, on the other hand, describe specific future individuals. These individuals will also, by virtue of being future people, share the properties, interests, and entitlements of their type yet differ in their identities and personal attributes. Whereas the existence of these individuals is affected by non-identity considerations – our actions are going to affect which individuals, or tokens, will later exist – the existence of future persons who will possess the general properties, interests, and entitlements attributable to future persons as such is not.

The distinction between the general properties of future people and the existence of particular future individuals is at the heart of Reiman's proposed solution to the non-identity problem. He differentiates between the general *properties* people possess – for example, the fundamental interest in what he calls normal functioning – and their existence as *particulars*. According to Reiman, the properties one enjoys weigh much heavier than one's existence as a certain particular in determining the quality of life one will live and are therefore at the basis of future people's moral entitlements. When we act in ways that adversely affect the interests and properties of future people, we violate their entitlements to meet the important interests they have by virtue of being future persons. That they, as certain particulars, would not have existed had we acted otherwise does affect the moral relevance of this violation (Reiman 2007: 83–87).

Kumar, too, bases his proposed solution on an important distinction between types and tokens. The way in which our actions can wrong future people is by violating what these persons are entitled to in terms of respect for their type. In other words, we have a duty to respect future people as future people and independently of the individual, or token, each future person will turn out to be. If our actions disrespect future people – for instance by violating their legitimate interests or basic entitlements – they frustrate this legitimate expectation of mutual respect and therefore our moral duties to future people. These moral duties are not linked to the existence of particular tokens of future people, that is, specific individuals, and therefore exist regardless of future people's identity (Kumar 2003: 111–115, 2015).

On my proposed account, as on Reiman's and Kumar's, acts do not need to make any future individual worse off in order to be unjust. In other words, it makes no difference that the same people will not exist whatever we do. Our duties are to those future people who will eventually come into existence, about whom we make presuppositions as specified but not individuated future persons. This means our obligations do not change regardless of whether our actions cause the set of future people A to exist or a different set of future people B. The same is true for the number of future people that will come to existence. What matters is that future people presupposed by our actions enjoy the basic requirements of justice, specified in this account as the conditions necessary to live an autonomous life. Our obligations are to ensure that our future-affecting actions respect these conditions, for example, by promoting or protecting the environmental circumstances that will allow future people to enjoy the conditions of autonomy. Because these duties are to future people as such, rather than to specific individuals, it makes no difference how many future people will exist, so long as they can at least satisfy their basic requirements of justice. In turn, the account may require us to act so as to bring less, yet better-off future people into existence, who will therefore be in a position to enjoy the conditions of autonomy, instead of more yet worse-off persons if these were to live lives below the threshold of autonomy.

By reframing our obligations in the proposed, identity-independent way, this approach to the scope of justice allows us to establish intergenerational duties of climate justice despite the fact that future people do not exist and that our actions are likely to affect their future identities. It therefore offers an alternative for grounding intergenerational duties of climate justice that can circumvent some important objections based on non-identity and non-existence considerations. While this is not meant to be the main selling point of the approach, it is an aspect which adds to its attractiveness as part of a theory of climate justice. As a successful account of the scope of climate justice the proposed account can therefore, in subsequent steps, serve as the basis for a detailed discussion of the substance of agents' obligations of climate justice. Yet even more importantly, it can, in the more immediate term, bolster the case for the relevant agents to take immediate climate action for future people.

Conclusion

This concludes my investigation into the appropriate scope of our duties of climate justice. The proposed account fits the framework constructed in previous chapters and explains that, and why, we have duties of climate justice to future people: it holds that agents acquire obligations of justice to future people whenever they engage in intergenerational actions, that is, actions which rely on presuppositions about future people. This chapter also clarified three important issues the account must be able to explain: the range of actions that are rightly seen as intergenerational, whether we can have obligations to both near and remote future people, and the role that intergenerational cooperation plays within the account.

I argued, first, that a very broad range of common, mundane actions are in fact intergenerational. Because all, or nearly all, collective agents relevant to climate change engage in at least some of these actions, the account gives rise to wideranging obligations of intergenerational climate justice. I then showed that this account can ground obligations to both near and remote future generations, since the temporal reach of the obligations depends on how far-reaching the presuppositions are which underlie the actions. The final issue I tackled was the role that intergenerational cooperation plays within the account of scope. I argued that cooperation indirectly affects the range of obligations which fall within the scope of the agent's duties, as cooperative actions necessarily rely on particularly specific presuppositions about future people, which, in turn, ground a broader range of obligations.

This account of the scope of climate justice successfully uses an action-centred, practical methodology to answer this book's central question. By assessing our actions, it shows that, and explains why, we have obligations of justice to future people despite the inherent uncertainty and complexity of climate change and the fact that our actions affect who is later born. This is an important step towards solving the broader intergenerational climate challenge but is also important in its own right: on the one hand, it provides additional theoretical tools to argue in favour of intergenerational duties of climate justice; on the other hand, it can be used in practice to buttress calls for immediate climate action for the sake of future generations.

Notes

1 See Chapter 1.
2 See Chapter 3.
3 See, for example, Cohen 2009; Elliot 1989; Kumar 2003; Parfit 1984; Velleman 2008; Woodward 1986, 1987. See also Chapters 1 and 5.

Reference list

Abizadeh, Arash. 2007. 'Cooperation, Pervasive Impact, and Coercion: On the Scope (Not Site) of Distributive Justice'. *Philosophy & Public Affairs* 35 (4): 318–358. https://doi.org/10.1111/j.1088-4963.2007.00116.x.

Broome, John. 1992. *Counting the Cost of Global Warming*. Cambridge: White Horse Press.

———. 2012. *Climate Matters: Ethics in a Warming World*. New York: W. W. Norton & Company.

Bundesministerium für Umwelt, Naturschutz und nukleare Sicherheit (BMU). 2016. 'Klimaschutzplan 2050'. www.bmu.de/fileadmin/Daten_BMU/Download_PDF/Klimaschutz/klimaschutzplan_2050_bf.pdf.

Caney, Simon. 2017. 'Human Rights, Responsibilities, and Climate Change'. In *Environmental Rights*, edited by Steve Vanderheiden. London: Routledge. https://doi.org/10.4324/9781315094427.

Cohen, Andrew I. 2009. 'Compensation for Historic Injustices: Completing the Boxill and Sher Argument'. *Philosophy & Public Affairs* 37 (1): 81–102. https://doi.org/10.1111/j.1088-4963.2008.01146.x.

Conca, James. 2015. 'Talking to the Future – Hey, There's Nuclear Waste Buried Here!' *Forbes*. 17 April. Online Edition. www.forbes.com/sites/jamesconca/2015/04/17/talking-to-the-future-hey-theres-nuclear-waste-buried-here/#2bca5ba92fef.

Deutscher Bundestag. 2019. *Bundes-Klimaschutzgesetz*. https://www.gesetze-im-internet.de/ksg/BJNR251310019.html.

Elliot, Robert. 1989. 'The Rights of Future People'. *Journal of Applied Philosophy* 6 (2): 159–170. https://doi.org/10.1111/j.1468-5930.1989.tb00388.x.

Evolution Library. 2001. 'The Current Mass Extinction'. *Evolution Library*. www.pbs.org/wgbh/evolution/library/03/2/l_032_04.html.

Feinberg, Joel. 1974. 'The Rights of Animals and Unborn Generations'. In *Philosophy & Environmental Crisis*, edited by William T. Blackstone, 43–68. Athens: The University of Georgia Press.

Howard, B.A., M.B. Crawford, D.A. Galson, and M.G. Marietta. 2000. 'Regulatory Basis for the Waste Isolation Pilot Plant Performance Assessment'. *U.S. Department of Energy Office of Scientific and Technical Information*. https://inis.iaea.org/collection/NCLCollectionStore/_Public/31/049/31049968.pdf.

Kumar, Rahul. 2003. 'Who Can Be Wronged?' *Philosophy and Public Affairs* 31 (2): 99–118. https://doi.org/10.1111/j.1088-4963.2003.00099.x.

———. 2015. 'Risking and Wronging'. *Philosophy & Public Affairs* 43 (1): 27–51. https://doi.org/10.1111/papa.12042.

Leconte, Jérémy, Francois Forget, Benjamin Charnay, Robin Wordsworth, and Alizée Pottier. 2013. 'Increased Insolation Threshold for Runaway Greenhouse Processes on Earth-Like Planets'. *Nature* 504 (7479): 268–271. https://doi.org/10.1038/nature12827.

Meyer, Lukas. 2016. 'Intergenerational Justice'. In *The Stanford Encyclopedia of Philosophy*, edited by Edward N. Zalta. Stanford: Metaphysics Research Lab, Stanford University. Summer.

O'Neill, Onora. 1996. *Towards Justice and Virtue: A Constructive Account of Practical Reasoning*. Cambridge: Cambridge University Press.

Parfit, Derek. 1984. *Reasons and Persons*. Oxford University Press.

Reiman, Jeffrey. 2007. 'Being Fair to Future People: The Non-Identity Problem in the Original Position'. *Philosophy & Public Affairs* 35 (1): 69–92. https://doi.org/10.1111/j.1088-4963.2007.00099.x.

Scheffler, Samuel. 2013. *Death and the Afterlife*. Edited by Niko Kolodny. Oxford: Oxford University Press.

Shiffrin, Seana Valentine. 2013. 'Preserving the Valued or Preserving Valuing?' In *Death and the Afterlife, by Samuel Scheffler*, edited by Niko Kolodny. Oxford: Oxford University Press.

Snyder-Beattie, Andrew E., Toby Ord, and Michael B. Bonsall. 2019. 'An Upper Bound for the Background Rate of Human Extinction'. *Scientific Reports* 9 (1): 11054. https://doi.org/10.1038/s41598-019-47540-7.

Stothard, Michael. 2016. 'Nuclear Waste: Keep Out for 100,000 Years'. *Financial Times*. 14 July. Online Edition. www.ft.com/content/db87c16c-4947-11e6-b387-64ab0a67014c.

United Nations Climate Change. 2015. 'Paris Agreement to the United Nations Framework Convention on Climate Change'. https://unfccc.int/sites/default/files/english_paris_agreement.pdf.

United States Nuclear Regulatory Commission. 2019. 'Backgrounder on Radioactive Waste'. 23 July. www.nrc.gov/reading-rm/doc-collections/fact-sheets/radwaste.html.

Velleman, J. David. 2008. 'The Identity Problem'. *Philosophy & Public Affairs* 36 (3): 221–244. https://doi.org/10.1111/j.1088-4963.2008.00139_1.x.

Woodward, James. 1986. 'The Non-Identity Problem'. *Ethics* 96 (4): 804–831. https://doi.org/10.1086/292801.

———. 1987. 'Reply to Parfit'. *Ethics* 97 (4): 800–816.

6   Changing perspective

The problem

We have known for some time that climate change threatens to cause great environmental damage that will substantially impact the lives of future people (see, for example, IPCC 1990). Starting from the Industrial Revolution, our greenhouse gas emissions have been dramatically changing the earth's climate, leading, for example, to worrying increases in ocean and surface temperatures. The most recent IPCC report paints a dire picture: already, anthropogenic greenhouse gas emissions since 1900 have caused average temperatures to rise by 1.1°C and ocean levels to increase by 0.2 metres, as well as leading to an unprecedented increase in precipitation, ocean warming, and the retreat of glaciers (IPCC 2021: 6). Anthropogenic changes to the climate have also meant that extreme weather events such as heatwaves, periods of high precipitation, and cyclones have become more frequent (IPCC 2021: 10–11). Unless we significantly reduce our greenhouse gas emissions and consistently maintain these more aggressive mitigation efforts throughout the next decades, by the end of the 21st century we will have exceeded not just 1.5°C degrees of warming but even the Paris Agreement's uppermost limit of 2°C. On the direst future emissions scenario considered by the Panel – which models the impact of our greenhouse gas emissions continuing to increase and doubling from current levels by 2050 – average global temperatures are predicted to increase by 3.3°C to 5.7°C by the end of the century. Even on its intermediate scenario, on which our emissions are kept constant at current levels until 2050 only to decline in the second half of the century, the IPCC predicts a warming of 2.1°C to 3.5°C by 2100. Their modelling further suggests that we will only be able to limit the average global temperature increase to between 1°C and 1.8°C if our emissions reach net zero by mid-century and, in addition, we employ CDR technologies thereafter (IPCC 2021: 15–17).

A direct consequence of such a sharp rise in temperatures is that large parts of the world could become inhabitable. In many other areas, increased heat is likely to put a significant strain on agriculture or the performance of daily activities. With greater warming, extreme weather events, too, will continue to become more frequent and more intense (IPCC 2021:19). All of this is going to severely threaten the health, safety, and livelihoods of future generations and directly affect

DOI: 10.4324/9781003258902-6

what future people are able to do and at what cost. Climate impacts are also very likely to lead to an increase in climate-induced migration as well as political, social, and economic instability. Both directly and indirectly, anthropogenic climate change is going to significantly limit the extent to which future people can enjoy the conditions necessary to live autonomous lives. The upshot of the IPCC's latest Assessment Report is that the outlook for future generations is bleak unless we significantly cut our emissions. In turn, this means that if we do reduce our greenhouse gas emissions within the next decade we can avoid the most harmful climate changes and preserve a safer environment for future generations. In other words, with more aggressive mitigation efforts now we can preserve an environment in which future people will be able to enjoy their conditions of autonomy.

After consecutive years of increasingly destructive weather extremes all over the globe, climate change is now receiving widespread attention from the media, civil society, and politicians. There is no doubt that interest in the impacts of climate change – of which we have known for decades – should have come much sooner. Yet the attention that climate change is receiving now suggests we may finally have reached enough momentum to make sustainable changes to our economies, which will benefit both ourselves and future generations. New campaigns and movements that have sprung up in the past few years suggest a growing awareness of the risks posed by climate change and of the urgent need to respond to them. Probably the best known among recent campaigns is the Fridays for Future movement founded in 2018. It was organised and is still spearheaded by young activists, mostly of school age, who are calling for climate justice and more aggressive action to reduce our greenhouse gas emissions. They are demanding that national governments listen to the available science and increase their mitigation efforts now in order to decrease the burden that younger generations, as they grow older, will have to bear if the changes in our climate go unchecked (Fridays for Future 2020).

Indeed, unless decisive action is taken now, future generations, including current younger generations, are going to be unfairly burdened on two fronts. On the one hand, the destructive impacts of climate change are going to continue progressing with time, harming not just the current younger generations but increasing the burden each subsequent generation will have to bear. On the other hand, if future generations want to ward off additional and even more destructive impacts of unchecked changes to our climate, they will have to take up a disproportionately large share of the mitigation burden to make up for the failings of present generations. This outlook is worrying, and it is clear to see why activists argue that present generations owe a duty of climate justice to the currently young to ramp up their mitigation efforts.

The case of unborn future generations, however, is unlike that of existing younger generations. Showing that the effects of climate change on non-existing future people are a matter of justice – and, consequently, that present generations have obligations of climate justice to non-existing future people – is not quite as straightforward. For one, the moral theories we commonly rely on to argue in favour of duties of justice do not apply easily to climate justice. Climate change is, as argued in Chapter 3, a particularly challenging moral problem and confronts our moral theories with various complicating factors. These problems include, among others, the temporal dispersion of causes and effects; the spatial and temporal dispersion of agency and vulnerability; the severity and uncertainty of future impacts; and the fact that due to its sheer scale, climate change and the underlying actions are going to affect the identity of future people. These issues are not only challenging by themselves but converge in a manner that is unique to climate change, making it a particularly complex problem of justice. All of this comes together to pose a particularly tough *intergenerational climate challenge* that puts existing moral theories to the test and opens intergenerational climate justice up to a range of objections. Arguing that our duties of climate justice ought, despite these objections and the challenges of complexity, include future generations within their scope is the first issue scholars of intergenerational climate justice face.

The second point of concern is that, more specifically, accounts of the scope of climate justice are often underdeveloped. That is, even where theorists of climate justice have assumed, despite these obstacles and objections, that our obligations are intergenerational in scope, their accounts have often insufficiently elaborated why this is so. Yet to solve the intergenerational challenge and respond to the most forceful objections we need a robust account of scope that can explain not just whether but why we owe climate justice to future people despite the intergenerational climate challenge. Given what is at stake, it is especially important that we frame our obligations to future people to mitigate the effects of climate change as a matter of justice. The impacts of climate change are going to threaten the lives of future people and severely restrict what they are able to do, with many of the predicted climate impacts limiting the extent to which future people will be able to live autonomous lives. Throughout this book, I have assumed that the conditions necessary to live an autonomous life are so fundamental that every agent has a justice-based entitlement to them and that, in turn, agents have an obligation to respect others' conditions of autonomy. That future conditions of autonomy are under threat emphasises the need to frame intergenerational climate change as a matter of justice. Moreover, obligations of justice are generally understood to be enforceable – contrary to duties of beneficence – and to take lexical priority over competing moral duties. Framing our intergenerational climate duties in terms of justice thus does two things: it highlights what is at stake and confers greater, much needed urgency to our obligations.

At the start of this investigation I set out to develop an account of the scope of climate justice that can overcome the obstacles posed by the intergenerational challenge and bolster existing accounts of intergenerational climate justice. My goal was to show whether, and give sound reasons why, we have obligations of climate justice to future people. Let me restate what I argued.

The argument

The framework

Throughout this book I argued that an account of the scope of climate justice must meet three criteria to successfully overcome the intergenerational climate challenge. As a first criterion, *an account of scope must be able to accommodate uncertainties about the future effects of our climate changing and mitigating activities (C1)*. Despite our certainty that anthropogenic emissions of greenhouse gases are causing the earth's climate to change, our ability to accurately predict which future climate changes will occur, and how exactly our actions are going to affect them, is still marred by substantial uncertainties. This criterion highlights that an account of the scope of climate justice must be able to respond to these unknowns. In Chapter 2, I argued that an account which takes our actions and their presuppositions as the locus of moral concern is better placed to do account for uncertainty than a consequentialist account. I also argued that in situations of uncertainty, the actions that an account of justice must focus on are those that impose *foreseeable risks*. I defined a risk as foreseeable whenever the agent (i) knows that her actions are interfering in the risk so as to increase its magnitude and (ii) has the knowledge and ability to disrupt this interference. Those actions are necessarily – whether or not the agent is aware of it – based on weighing the preferences of the risk-imposing agents against those of risk-exposed future people. This, in turn, means that those actions rely on presuppositions about the capabilities and vulnerabilities of future people. This brings out one of the key elements of my account of scope: I argued that whenever actions are based on presuppositions about others, the agents must include these others within the scope of their obligations of justice.

As a second criterion, I argued that *an account of the scope of climate justice must be able to respond to climate change as a complex problem of justice (C2)*. As highlighted by the intergenerational challenge, climate change consists of a number of individually tough problems. In addition, these issues converge to make climate change a particularly complex moral problem, so much so that existing moral theories seem unfit to deal with climate change in its entirety. In Chapter 3, I argued that a successful account of scope must be able to overcome the complexity created by the convergence of these individual issues. I suggested that one promising way in which we can respond to the challenge of complexity is by opening up our moral theories to methods that are new to the domain of climate justice and by using these methods to repurpose the moral values we already hold. This is precisely what the action-centred methodology proposed in Chapter 2 allows us to do. In Chapter 3 I expanded the arguments of Chapter 2 and suggested that – in addition to actions imposing foreseeable risks – we engage in a variety of other activities that rely on presuppositions about future people. I called these *intergenerational actions* and argued that agents who engage in them acquire an obligation to extend the basic requirements of justice to those future others whom their actions presuppose. These basic requirements of justice are to respect future people as agents and to respect their conditions of autonomy. In other words, I argued that agents have an obligation to include others on whom their actions rely into the scope of their obligations of justice and consequently of climate justice. This includes intergenerational actions, so that agents have an obligation to include future people in the scope of their duties of justice whenever they engage in intergenerational actions. Some intergenerational actions – strong intergenerational actions – are cooperative in nature; that is, they require future people to contribute to be completed successfully. I argued that these actions ground a form of intergenerational cooperation, which highlights the deep intergenerational connections that exist throughout our societies.

In Chapter 4 I then argued that, as a final criterion, *an account of the scope of climate justice must include each person for their own sake and as an end in themselves (C3)*. This criterion determines how future people ought to be included in the scope of justice, thus setting important ground rules for any substantive account of climate justice that may be built on this account of scope. Meeting this requirement ensures that the account does not sanction treating future people as mere means and that it does not justify unjust future worlds. I argued that we run a real risk of doing so if we include future people in the scope of our obligations because of our dependence on them and that we must instead include them for their own sake.

The account

Based on this framework, I formulated an account of the scope of climate justice that centres on the concept of intergenerational actions and the resulting obligations. The account is summarised in Chapter 5. Before proposing the final version of the account, in this chapter I also clarified three final issues: I showed that a very broad range of actions we regularly engage in are in fact intergenerational because they, too, rely on presuppositions about the continued existence of a chain of future humans; that the account allows us to include remote and near future generations within the scope of climate justice, depending on the presuppositions on which our actions are based; and that, compared to non-cooperative actions, cooperative intergenerational actions ground particularly extensive obligations to future people because they necessarily rely on more detailed presuppositions about them. With these issues in mind, the final account of the scope of climate justice I proposed is as follows:

Account (A2)

1   Our actions are based on certain presuppositions about others, including, often, about future people; these presuppositions determine whether and to whom we owe duties of justice and climate justice.
2   If our actions rely on presuppositions about future others, we must also recognise these future people as independent moral agents in our underlying principles of justice (*coherence requirement*) and extend the basic requirements of justice to them (*normative requirement*).
3   The scope of an agent's duties of justice therefore extends to all those present and future others whom their actions presuppose. If an agent is engaged in activities relevant to climate justice, the scope of their duties of climate justice also extends to all others whom their actions presuppose.
4   The scope of an agent's obligations can extend to near and remote future generations, depending on the presuppositions their actions are based on.

5   The kinds of obligations that fall into the scope of an agent's duties – whether they include only a duty to respect future people's basic conditions of autonomy or extend to more specific capabilities – depend on the specificity of the presuppositions on which the agent's actions are based.

Changing our perspective

The aim of this book has been to answer one main question: do we have obligations of climate justice to future people, and if so, why? In response, the proposed account of scope shows that agents owe duties of justice to future people whenever their actions are based on presuppositions about those future others. This approach to scope couples an action-centred methodology with the requirement to respect the values of agency and autonomy. On the one hand, this way of repurposing existing moral concepts with a novel methodology allows the account to overcome the intergenerational climate challenge and show that future people fall within the scope of our obligations of climate justice. On the other hand, this approach to the scope of climate justice is unique in that it also succeeds in reframing climate change as an intrinsically intergenerational problem. This intergenerational framing emphasises that climate change differs from most problems of justice we know and that it must therefore be met with equally distinct theoretical and practical responses in order to be tackled successfully.

The proposed, reframed approach to the scope of climate justice has four important implications for the theory and practice of climate justice. First, it puts us in a position to respond to the theoretical issues I had set out to overcome at the start of this project. That is, it allows us to respond to the non-identity problem, the non-existence challenge, and address issues of complexity, as well as providing us with novel philosophical arguments in favour of extending the scope of our obligations. Second, it allows us to reassess our current climate policies in light of our duty to respect the autonomy conditions of future people. Thirdly, an understanding of climate change as a deeply intergenerational phenomenon may motivate us and allow us to push more forcefully for greater climate action by changing the way our climate changing and mitigating activities are seen within their societal context. Lastly, the intergenerationality of our actions and our resulting duties to future generations give us grounds on which to reassess the legitimacy and shape of our political institutions. Let me elaborate each of these implications in turn.

Theoretical issues

We can use the proposed account of scope to respond to theorists who deny the very possibility of intergenerational climate justice. As argued in Chapter 5, this action-centred account does not rely on a notion of harm to ground intergenerational obligations nor does its link our duties of intergenerational justice to the identity of specific future persons. Because of this, it does not fall victim to the non-identity problem. It therefore allows us to convincingly argue for intergenerational obligations of climate justice despite the fact that climate change is world-constituting, that is, that most of our actions related to climate change affect who will later be born. In addition, it does not require the interests it protects – future people's interests in the ability to live an autonomous life – to exist at the same time as our obligations to respect these interests and the entitlements that follow from them. For this reason, it also allows us to circumvent the non-existence challenge to intergenerational justice.

At the same time, because the proposed account reframes the grounds of our intergenerational obligations and clearly links our duties to our own agency it provides us with novel arguments in support of existing accounts of intergenerational climate justice. An action-centred methodology is especially helpful when we are dealing with complex and uncertain moral problems like climate change. Because the proposed account ties our duties to our own actions and their presuppositions, we only need to rely on information that is easily accessible to us even in complex, uncertain situations like climate change to determine the scope of our duties. Using an action-centred methodology we can therefore establish duties of intergenerational justice despite the complexity of climate change as a moral problem. In addition, as argued in Chapter 1, earlier accounts of intergenerational climate justice have often lacked a sufficiently clear and forceful defence of their claim that climate justice ought to be intergenerational. This has made those accounts vulnerable to objections like the non-identity problem. The account of scope developed in this book provides a more robust theoretical foundation on which to ground those already existing accounts of the substance of climate justice. Thus, proponents of intergenerational climate justice can also now resort to the concept of intergenerational actions, the coherence requirement, and the normative requirement I defended in earlier chapters to argue more decisively in favour of intergenerational obligations.

Rethinking our climate policies

Establishing a duty to respect the conditions of autonomy of future people, as I have done in this book, has significant practical implications relating to how we ought to formulate our climate policies. According to the account of scope I have proposed, agents acquire duties of justice to future people whenever they engage in actions that rely on presuppositions about them, in other words whenever they engage in intergenerational actions. Intergenerational actions abound in our societies: they include many climate plans and policies, long-term construction projects, or the practice of public debt. Because intergenerational actions are so common, we can assume most, if not all, collective agents relevant to climate justice to be engaged in one or more action that relies on presuppositions about future people. To recall, agents relevant to climate justice are agents who engage in actions that directly affect the extent to which we are able to mitigate, adapt, or otherwise respond to anthropogenic climate change. This includes local, regional, and national governments, many intergenerational organisations, civil society organisations, and private industry. In turn, these agents acquire an obligation to act in ways that respect the autonomy conditions of future people, including whenever they engage in activities relevant to climate justice.

Once we have established that these duties exist, and that they are owed by present collective agents to future generations, we have a new criterion for assessing and reformulating our climate policies. That is, we now must evaluate whether our climate policies respect the autonomy conditions of future generations. I argued in Chapter 2 that no matter which substantive account of autonomy we adopt, the conditions necessary to live an autonomous life are going to include things like life, bodily integrity, freedom of movement, shelter, education, nutrition, and health care. Many have already temporarily or permanently been rendered unable to enjoy one or more of these conditions due to the effects of climate change, most notably because of the increased intensity and frequency of extreme weather events. In its most recent report, the IPCC reaffirms that the negative impacts of climate change we have been seeing – including extreme weather events – will intensify if temperatures continue to rise (IPCC 2021). This means that continued warming will lead to even more people suffering the effects of extreme weather events, being displaced from their homes by rising sea levels or losing their livelihoods to desertification. In turn, an increasing number of people are going to be deprived of a chance to enjoy some, or even all, conditions necessary to live an autonomous life.

To comply with our intergenerational duties and avoid as many future violations of autonomy conditions as possible we ought to limit future warming as much as we can and to the extent that we can do so without endangering our own autonomy conditions. Although there is no such thing as a safe level of warming, a good benchmark is the target of maximum 2°C, but aiming at 1.5°C, set out in the Paris Agreement (IPCC 2018, 2021; United Nations Climate Change 2015). This target takes into account the need to limit the most dangerous effects of future warming, while considering the temperature rise that has already been locked in by past emissions. Moreover, it requires reasonable mitigation efforts that present generations can shoulder without endangering our own conditions of autonomy. Current climate policies, however, are far off this target: according to the Climate Action Tracker the mitigation policies pursued at the time of writing put us on track for a warming of 2.7°C by 2100. If governments around the globe were to adhere to their current pledges and targets, this would bring the average temperature increase by the end of the century to a slightly cooler 2.1°C (Climate Analytics and New Climate Institute 2021). Both predictions are far off the safer target of 1.5°C and miss even the upper bound of 2°C. The upshot is that current climate policies are failing to meet our obligations to future people by failing to preserve, or in many cases actively destroying, an environment that would enable future generations to safely enjoy the conditions necessary to live an autonomous life.

That our mitigation efforts are failing to meet our intergenerational obligations is now starting to be picked up by the courts too. In a widely discussed judgement, the German Constitutional Court ruled in March 2021 that the mitigation efforts pursued by the German government were insufficient to protect the liberties of future people. It held that, unless the government significantly ramped up its mitigation efforts, future generations of German citizens would be left with too high a mitigation burden for them to enjoy the full range of liberties to which they are entitled (Bundesverfassungsgericht 2021). This ruling reaffirmed the main argument of this book and brought future people's entitlement to fundamental freedoms to the centre of the public discourse on climate change. It confirmed that unless we tighten our climate policies to ensure we meet our temperature targets we will leave behind an environment in which future people will be unable to enjoy their conditions of autonomy.

If we continue on the current path, our greenhouse gas emissions will contribute to creating a more dangerous environment for future people. This, in turn, will affect future conditions of autonomy in two ways. On the one hand, a hotter climate is going to pose a direct danger to the autonomy of future people. As higher temperatures or rising sea levels render parts of the world inhabitable, and as extreme weather events become more frequent and intense, it is very likely that many future people will be unable to enjoy some or all conditions of autonomy. On the other hand, if they want to preserve as safe an environment as possible, future generations will have to try much harder to limit these impacts by picking up a disproportionately large share of the mitigation burden. I argued in Chapter 4 that this – present generations using too much of the remaining carbon budget and letting future generations pick up their slack – amounts to future people being used as mere means. It also means that – if they want to avoid the most dangerous climate changes – future generations will have to use a disproportionately large share of their resources to combat climate change. In turn, they will not be able to use these resources elsewhere and, more generally, to choose autonomously what to use their time and money for. This in itself is going to restrict future generations' conditions of autonomy. It also makes it more likely that future people will suffer additional restrictions to autonomy in those other, underfunded policy areas. The upshot is that we must drastically change our behaviour to meet our intergenerational obligations of climate justice. In other words, to comply with the duties established by this account we must reassess and tighten our climate policies.

Tackling climate change as an intergenerational problem

In addition to grounding obligations of justice, the intergenerationality of our actions also indicates that climate change is a deeply and structurally intergenerational phenomenon. This affects how we ought to view and tackle it as a moral and practical problem. While it has long been evident that current greenhouse gas emissions are going to have severe impacts on even remote future generations, I do not believe that the true scope of the intergenerationality of climate change has been sufficiently emphasised so far.

Throughout this book, I showed that many of the actions we take in response to climate change are intergenerational; that is, they are based on presuppositions about future people. Some examples are the Paris Agreement and local and national climate plans. Equally, many of the emitting activities that are causing climate change are intergenerational. Take, for example, long-term construction projects. I argued that long-term projects which require the work of more than one generation are examples of strong, cooperative intergenerational actions. All such long-term, large-scale construction activities will emit large amounts of greenhouse gases during the construction work itself as well as for the production and transport of the necessary raw materials, among others. Their overall emissions will increase further still if what is being built is itself highly polluting or encourages polluting behaviour such as a coal plant, mine, or airport. Similarly, consider long-term public investments or public debt, both of which, too, I argued are intergenerational actions. If those investments are used to fund highly emitting industries – for examples in the fossil fuel sector – or the newly borrowed funds are reinvested in unsustainable, polluting practices, then both these actions, too, will contribute to increasing the overall rate of climate change.

Examples like these suggest that many present activities that are causing climate change, as well as many actions we take in response to it, rely on a subconscious and deeply ingrained understanding of our societies as intergenerational. This intergenerationality enables the connections based on presuppositions, expectations, and trust on which our climate changing and mitigating intergenerational actions are based. To summarise: on the one hand, we know that climate change is caused and can only be tackled by our own actions; on the other hand, these activities are embedded in an intergenerational understanding of society and deeply linked to future people via the presuppositions on which they rely. This paints climate change as a problem that can only be adequately understood if we take into account not just its descriptive, scientific intergenerational element – given by the temporal dispersion of its causes and effects – but also remember to consider its normative, social, and moral intergenerational elements. While the former is widely acknowledged – with the exception, possibly, of climate change deniers – I believe the latter have, until now, been insufficiently emphasised.

Acknowledging that climate change is in this way an intrinsically intergenerational problem has two important implications. On the one hand, this assessment of climate change as deeply intergenerational should change the way in which we understand and view the problem as a whole. It suggests that a successful response to climate change requires us to shift our perspective on the problem to reflect its intergenerational nature, for example, by adapting the moral, political, and discursive tools we apply to it. One of the ways in which we must change our approach to climate change is by adapting our policies to reflect the intergenerationality of our actions and our corresponding duties.

On the other hand, recognising that climate change is an intrinsically intergenerational problem has the potential to motivate a more aggressive and active response to it, as it places our actions within an interconnected, intergenerational context. Emphasising the connections between our own agency and future people can potentially decrease the psychological distance between us and future generations – especially more distant ones – and by doing so increase our willingness and motivation to discharge our intergenerational obligations. At the same time, this intergenerational reframing of climate change spells out the extent to which we ourselves rely on future people – for example, for the completion of our intergenerational actions – and can in this way motivate action on more self-interested grounds, too. An awareness of climate change as intrinsically intergenerational, combined with the realisation that we owe duties of climate justice to future people, and an understanding of what our climate policies should look like to comply with our obligations, can thus change the way our mitigating and climate changing activities are perceived and motivate us to greater climate action. Samuel Scheffler, in his striking account of our relationship to the future of humanity, argues that we, present generations, are in fact dependent on future people (Scheffler 2013, for example, pages 78–79). The existence and prevalence of intergenerational actions that I have defended show one way in which this is true. As I have argued in this chapter, it is true even for our actions surrounding climate change – a problem which, paradoxically, is also posing one of the greatest threats to future people and their conditions of autonomy.

The account of scope I have defended shows that we have sound theoretical grounds on which to justify obligations of climate justice to future people; in addition, pointing to the deep intergenerational connections that exist within our societies and which are at the heart of both climate change and our responses to it can and should motivate us to act on these duties. At the very least, it gives us a new perspective from which to argue for more aggressive action, as well as additional reasons in favour of acting accordingly.

Intergenerational actions and democratic institutions

On a more fundamental level, acknowledging that climate change is a deeply intergenerational problem should prompt us to rethink the institutions we rely on to legislate for and enforce climate justice. The present-centred workings of democratic institutions – including national governments and parliaments as well as international and intergovernmental organisations – stand in stark contrast to the prevalence of intergenerational actions in our societies. Political institutions can often display wrongful short-termism; that is, they prioritise short-term benefits over the interests of future generations (González-Ricoy and Gosseries 2016b; MacKenzie 2016). This is especially concerning when dealing with long-term problems like climate change as the only way to limit dangerous climate changes is for present generations to take costly mitigation measures that will benefit mostly future people. The prioritisation of present benefits over future interests is one of the reasons why present generations have been passing the buck of climate action to future generation.

The prevalence of short-termism in democratic institutions stands out against the findings of this book. I suggested, for one, that climate change is a deeply intergenerational problem. Many of our actions, including actions relevant to climate justice, are intergenerational and, in other words, rely on presuppositions about the existence and capabilities of future generations. Some of these actions are cooperative intergenerational actions, meaning that they cannot be completed successfully unless future people contribute to them. The upshot is that, through our presuppositions, future people become an integral part of our actions and, with them, of our societies. For another, I argued that by engaging in intergenerational actions we acquire intergenerational obligations of climate justice to future people that require us to respect their fundamental interests in the conditions of autonomy.

These arguments raise two sets of questions for our institutions. The first concerns the enforcement of our intergenerational duties. Do we need to adapt the ways in which political institutions function if we want our duties to future people to be respected? Complying with our obligations requires us to give much greater consideration to future interests versus present benefits, contradicting the short-termism currently displayed by many democratic institutions. The second set of questions concerns the structure of our institutions. Does the prevalence of intergenerational actions give us reasons to change the make-up of institutions to ensure that all those involved in our actions – including future people – are also included in the process of legislating for those activities?

Some countries have already experimented with future-oriented institutions. In 1993, Finland set up a Committee for the Future that still serves as a standing committee in the Finnish Parliament and which aims to create a discourse with the government on future problems and opportunities (Parliament of Finland 2021). Other well-known, though more short-lived, examples are Israel's Commissioner for Future Generations, which was set up in 2001 and discontinued in 2006, and Hungary's Ombudsman for Future Generations, a post that existed from 2008 until 2012 (Read 2012; United Nations Secretary General 2013). Fuelled partly by the increasing pressure to act on intergenerational problems like climate change, the academic discourse on the need for and opportunities provided by future-oriented institutions has been growing too (see Boston 2014; Read 2012; all contributions to Gonzáles-Ricoy and Gosseries 2016a; Tremmel 2006).

The account of scope proposed in this book suggests that future work – both academic and political – should push further in this area. Our engagement in intergenerational actions and the resulting duties contribute to the attractiveness of future-oriented institutions as tools that may allow us to better comply with our intergenerational duties and, more generally, as a way to align the institutional foundations of our societies with their intrinsically intergenerational nature. More specifically, future work can rely on the arguments of this book to explore the possibility of developing legislative and executive institutions to be more in line with the intergenerational nature of societies. On the basis of the proposed account we may want to consider, on the one hand, whether legislative branches of governments ought to be expanded to include some form of representation of future generations in addition to the present electorate. This extension would ensure that all agents who are part of intergenerational actions are represented within the body that formulates the laws and policies that result from these actions – in other words that formalises the intergenerational duties that intergenerational actions ground.

On the other hand, the fact that we have intergenerational duties of justice, including climate justice, suggests that we will need executive branches of government to ensure these obligations are enforced. Extending the powers of executive institutions to explicitly include the enforcement of intergenerational duties would ensure that we do not violate our intergenerational duties as we engage in intergenerational actions. The ruling on climate justice by the German Constitutional Court showed that the executive is, to some extent, already empowered to enforce intergenerational duties, at least in Germany. But it also emphasised that, for now, governments are failing to exercise this power unless explicitly prompted to do so by the courts. Important questions, for example, relating to the democratic legitimacy of future-oriented institutions, will need to be answered before a more detailed proposal for institutional change can take shape and eventually be implemented. Nevertheless, the account of scope proposed in this book suggests that this is an important avenue we ought to continue exploring to ensure our climate policies and the institutions that govern them are intergenerationally just.

Where to go from here

A sound account of scope is important in its own right, as it can justify our obligation to bear some of the costs of climate change and ensure future people are not arbitrarily disadvantaged by our relative position of (temporal) power over them. At the same time, it is only the start of what needs to be a much broader examination of climate justice, its foundations, and its implementation. Since any robust theory of climate justice must begin with a strong account of scope to successfully meet the challenges posed by climate change as a moral problem, future work can rely on my proposed account of scope as the basis to formulate an account of what, and how much of it, we owe future people. To arrive at a complete theory of climate justice, additional questions would need to be answered. An important one, for example, is how conflicts between competing duties are to be resolved. Such conflicts could arise between duties of intragenerational and intergenerational justice, between various intergenerational duties, or between duties of climate justice and other domains of justice. The proposed account of scope will be of use as it can establish an initial set of obligation to serve as the basis for an account of climate justice that deals with these questions.

This investigation, while focused on the intergenerational scope of climate justice, has itself raised important substantive issues that can spark interesting future discussions. In Chapter 3, for example, I argued that one way to respect the future people whom our actions expose to risks is to give due consideration to their conditions of autonomy when engaging in risk-imposing activity. In Chapter 4, I then argued that future people ought to be included in the scope of justice for their own sake. This requirement has a significant impact on the substantive principles the account will allow, limiting the permitted rules to those that respect future people as ends in themselves and autonomous agents. Both points were prompted by and are closely linked to questions about the scope of our obligations yet raise issues that can only be dealt comprehensively with the help of a more complete, substantive theory of intergenerational climate justice. This makes them promising starting points for future investigations. More generally, the account's focus on agency and the conditions of autonomy sets clear boundaries for substantive accounts of intergenerational climate justice that may be built on it, as it implies that the minimum requirement of justice must be the protection of agents' fundamental needs or conditions of autonomy. It is now for substantive accounts of justice to spell out the fundamental conditions of autonomy and to devise principles that best protect the entitlements of those who fall within the scope of justice. It should be noted, however, that what I have argued in this book does not rule out the existence of additional moral obligations alongside those grounded by the proposed account. Whether we have additional intergenerational obligations, and if so, what they are, is a further question to be tackled by future work.

Notably, the arguments put forward in this book as well as the resulting account of scope may also be used to fix the scope of our obligations in other domains of justice. On the one hand, arguments related to agents relevant to climate justice have implications for other areas of justice too. Agents relevant to climate justice acquire intergenerational obligations by engaging in intergenerational actions, and these obligations will apply to their actions relevant to climate justice as well as to any other activities they engage in. In other words, agents relevant to climate justice that engage in intergenerational actions acquire an obligation to respect the autonomy conditions of future people in all intergenerational actions they engage in, whether they are related to climate justice or not.

On the other hand, the arguments I presented also lend themselves to be applied, mutatis mutandis, to any other sets of agents besides those relevant to climate justice. Where these agents are engaged in intergenerational actions of any kind, they too will acquire intergenerational obligations of justice. These obligations will apply to all intergenerational actions they are engaged in. Depending on the agents it is applied to, the proposed account of scope can therefore serve to fix the scope of our obligations well beyond the domain of climate justice.

New arguments in favour of intergenerational obligations of climate justice, such as the ones proposed in this book, can and should also play a role in civil society movements advocating for increased climate action. I have already argued that understanding climate change as an intergenerational problem has the potential to motivate agents to engage in more aggressive climate action. Current climate justice movements – for example, the Fridays for Future movement – could capitalise on this potential. An appeal to the intergenerationality of our actions, the deeply intergenerational nature of climate change, and our reliance on presuppositions about future people could be used, for example, to bolster the campaign's existing arguments for climate justice and climate action. Clearly, such references and appeals are not going to be as a silver bullet to wake us from our climate lethargy. Yet a change of perspective can often work wonders, and, given what is at stake, we have no excuse not to try.

Reference list

Boston, Jonathan. 2014. *Governing for the Future: Designing Democratic Institutions for a Better Tomorrow*. Bingley: Emerald.

Bundesverfassungsgericht. 2021. 'Constitutional Complaints Against the Federal Climate Change Act Partially Successful'. 29 April. www.bundesverfassungsgericht.de/SharedDocs/Pressemitteilungen/EN/2021/bvg21-031.html.

Climate Analytics and New Climate Institute. 2021. 'The CAT Thermometer'. *Climate Action Tracker*. May. https://climateactiontracker.org/global/cat-thermometer/.

Fridays for Future. 2020. 'Fridays for Future'. April. https://fridaysforfuture.org.

González-Ricoy, Iñigo, and Axel Gosseries. 2016a. *Institutions for Future Generations*. New York: Oxford University Press.

————. 2016b. 'Designing Institutions for Future Generations'. In *Institutions for Future Generations*, edited by Iñigo González-Ricoy and Axel Gosseries, 3–23. Oxford: Oxford University Press. https://doi.org/10.1093/acprof:oso/9780198746959.003.0001.

IPCC. 1990. *Climate Change: The IPCC Impacts Assessment*. Canberra: IPCC.

————. 2018. *IPCC Special Report Global Warming of 1.5°C*. Incheon, Republic of Korea: IPCC.

————. 2021. *Climate Change 2021: The Physical Science Basis. Contribution of Working Group I to the Sixth Assessment Report of the Intergovernmental Panel on Climate Change*. Cambridge: IPCC.

MacKenzie, Michael K. 2016. 'Institutional Design and Sources of Short-Termism'. In *Institutions for Future Generations*, edited by Iñigo González-Ricoy and Axel Gosseries, 24–46. Oxford: Oxford University Press. https://doi.org/10.1093/acprof:oso/9780198746959.003.0002.

Parliament of Finland. 2021. 'Committee for the Future'. www.eduskunta.fi/EN/valiokunnat/tulevaisuusvaliokunta/Pages/default.aspx.

Read, Rupert. 2012. *Guardians of the Future: A Constitutional Case for Representing and Protecting Future People*. Weymouth: Green House.

Scheffler, Samuel. 2013. *Death and the Afterlife*. Edited by Niko Kolodny. Oxford: Oxford University Press.

Tremmel, Jörg, ed. 2006. *Handbook of Intergenerational Justice*. Cheltenham and Northampton, MA: Edward Elgar.

United Nations Climate Change. 2015. 'Paris Agreement to the United Nations Framework Convention on Climate Change'. https://unfccc.int/sites/default/files/english_paris_agreement.pdf.

United Nations Secretary General. 2013. *Intergenerational Solidarity and the Needs of Future Generations*. New York: United Nations.

Index

action: against climate change 11, 93; about risk 36–37; related to climate justice 29, 31, 42, 64–65, 68, 100

action-centred methodology, for climate justice 59–73, 74; collective agents, intentions, and assumptions 59–61; intergenerational actions 62–67; intergenerational obligations 68–70; justice, uncertainty, and complexity 59; presuppositions behind organised collective actions 61–62; scrutinising the presuppositions behind collective actions 70–73

adaptation plan, for a coastal area 12, 111, 114

afterlife: care and the justification of justice 94–95; caring about 91–94; collective 82, 86, 89; importance of 91–95; *see also* collective afterlife; future persons

afterlife conjecture 98, 102–103, 106–107, 114; activities affected by 109; purposes of 108; scope of climate justice 105–106

agents, intergenerational actions of 101

agents relevant to climate justice: conditions of autonomy 38–39, 45, 124; contributions and vulnerability of 56; duties of justice 114, 128; future-oriented action under deep uncertainty 72; interactions with artificial intelligence 116; intergenerational actions 66; intergenerational obligations 72; justice-based entitlement 124; moral philosophy of 69; obligations of climate justice 100, 106, 110, 114,

119; organised collective agents 45, 69; presuppositions behind the actions of 70–73; risk-imposing 100; theory of collective responsibility 71

Andina, Tiziana 63–64

anthropogenic climate change 3, 7, 12, 60, 100, 123, 128

anthropogenic emissions of greenhouse gases, impact on climate 1, 45, 125

artificial intelligence 116

asteroid: collision, risk of 25, 45; doomsday 24–25, 31–35, 47; protection, of future generations 25

asymmetric vulnerabilities, to climate change 56, 73, 101

autonomy: of future people 80, 88, 128, 130; duty to respect other as agents 39–42; impacts of climate change on 39; view of justice 7, 38

autonomy conditions 39, 45–46, 69, 95, 117; of autonomous agents 16; and duties of justice 7; duty to respect 44; foreseeability condition 7; for foreseeable risk-imposition 38–39; of future people 40–41, 117, 127–129, 135; impacts of climate change on 39; importance of respecting 49; justice-based claim 38; obligation to respect 80, 110–111; risk-exposed 39; violations of 40, 129

Barry, Melissa 62

beneficence 3–12, 25, 124

Broome, John 4–5, 13; account of climate ethics 19n3; distinction between public and private morality 19n5

burning coal, impact of 93
business as usual (BAU) 27,
    49n7

cancer research 82, 92–94
Caney, Simon: on obligations to future
    people 9; recipient-based view of justice
    8, 115
carbon budget 85–86, 130
carbon dioxide removal (CDR)
    technologies 40, 67
carbon neutral economy 65
challenges of intergenerational climate
    change 57–58, 90, 94, 124
Children of Men, The (James) 81
civil society groups 60
claim of justice, against risk-imposition 39
classic utilitarianism, theory of 28–30
climate: ethics 4, 11, 19n3, 19n5, 55;
    migration induced by 123
Climate Action Plan 2050 (Germany)
    64–65, 72, 107
Climate Action Tracker 129
climate change 93; ability to predict 46;
    agents' contributions and vulnerability
    to 56; anthropogenic 3, 60, 100, 128;
    challenges of 57–58; complexity of 3,
    17, 56; consequentialist approach to
    28–30; empirical overview of 27–28;
    ethical response to 9; future effects of
    1; and future generations 1–3; Gardiner
    hypothesises to 9; harmful effects of
    86; impact of greenhouse gas emissions
    on 1; intergenerational dimension of
    1; intergenerational nature of 74; as
    intergenerational problem 130–132;
    loss of livelihood caused by 3; as a
    matter of intergenerational justice 6–10;
    mitigation of 66–67, 85; moral norms
    needed to respond to 101; New York
    climate change plan 65; requirements
    of 13; theoretical overview of 26–27;
    uncertainty predictions in 26
climate justice 56, 74, 99; actions related
    to 64–65; agent's obligations of 100;
    and beneficence 3–12; defined 2, 6;
    demands of 58; discounting for time
    10–12; duties of 2, 69; and future

climate changes 24–26; and non-
    existence challenge 5–6; non-identity
    problem 4–5; obligations of 2, 5, 9, 12,
    16; of organised collective agents 45;
    overinclusiveness of 30–32; principles
    of 3, 6, 16, 94; pure time preference 10;
    recipient-based account of 115; struggle,
    complexity of 58; traditional concepts
    of 58; underinclusiveness of 32–35;
    Valentini's autonomy view of 38,
    50n11; *see also* duties of climate justice;
    intergenerational climate justice; scope
    of climate justice
climate policies: cost-benefit analyses
    of 29; criterion for assessing and
    reformulating 129; formulation of 128;
    rethinking 128–130
cluelessness 30
collective action: definition of 70;
    presuppositions behind 70–73
collective afterlife 95n1; existence of 82;
    loss of 103; need for 108; notion of 81;
    prospect of 82
collective agency 70
collective agents 13, 59–61, 65, 69–73,
    105–106, 114, 119, 128–129; defined
    75n5; non-identity problem 4–5;
    organised 12, 19n5, 45; relevant to
    climate justice 62, 75, 93
collective intentions 70, 75n5, 105
collective responsibility, agential theory of
    69, 71
collective will, formation of 71
community of human valuing 89
complexity of climate change 70, 101,
    119; dealing with 101–102; overcoming
    73–74; problem of 53–57, 58, 73
consequentialism 29, 30, 35, 46, 47,
corporate identity 60
cost-benefit analyses, of climate policies
    9, 11, 29

debt restitution 61
decision-making: capacities and
    capabilities 60; climate-political 29;
    notion of reasonableness as tool for 45
Deep Ecology movement 54
deforestation, impact of 93

desertification, due to climate
  change 129
discounting for time 8, 10–12
duties of climate justice 7, 16, 32, 42, 47,
  60, 98–99, 113, 124; action-based 62;
  and foreseeability of risk of climate
  change 47; to future people 58, 59, 128;
  intergenerational 114; moral evaluation
  and grounds of 49

electricity production, from renewable
  sources 65
emergent risks 28
emission reduction goals 65
environmental damage, caused by climate
  change 122
environmental degradation 85, 87, 93
Erskine, Toni 60
European Union (EU) 106
evaluative dependency thesis 96n3

financial planning 64
foreseeability, of risk of climate change:
  concept of 36–38, 44, 46; conditions
  of 46–47; consequentialism of 46–48;
  defined 44, 46; duties of justice and 47;
  moral rightness of 47; notion of 100;
  requirement for 43; understanding of
  42–46
fossil fuel: economy 27; investments in
  sector of 74, 94; technologies 40
freedom of movement 38–39, 129
French, Peter 60
Fridays for Future movement 123, 135
future generations, scope of (climate)
  justice for 106–109; near future
  generations 107; remote future
  generations 107–109; unborn future
  generations 123
future-oriented projects 82, 92–93
future-oriented review mechanism 67
future-oriented temperature goal 67
future persons: autonomous life 88;
  capabilities and vulnerabilities of 106;
  climate justice and criterion of inclusion
  of 102; collective afterlife of 82, 86, 89;
  costs of present benefits to 93; effects
  of climate change on non-existing 123;

ends and means of 84–89; equal moral
  value of 91; game analogy 89–91;
  implications for intergenerational justice
  87; importance of the afterlife 91–95;
  moral value of 80–84; notion of 108;
  obligations of intergenerational justice
  to 73, 89, 124; "ongoing enterprise" of
  humanity 87; practice of valuing 87;
  Scheffler's account of existence of 84;
  value-laden lives 82; vulnerability of 90;
  see also afterlife; presuppositions, about
  future people

game analogy 89–91
Gardiner, Stephen 3, 9, 17–18, 53, 55–58,
  73, 87, 101
German Constitutional Court 129, 133
glaciers, retreat of 122
global justice 54
global mean surface temperature, rise in 28
Goodin, Robert 60
Greaves, Hilary 30–31
greenhouse gas emissions 33, 131;
  anthropogenic 1, 27, 45, 122; business
  as usual (BAU) scenario 27; campaign
  for reduction of 123; change of the
  earth's climate due to 1, 85, 122; CO_2-
  equivalent concentration of 76n9; costs
  of 27; New York's climate change plan
  to reduce 65

Homo Sapiens, lifespan of 108
humanity, extinction of 81, 108
human rights: approach to
  intergenerational climate justice 8; of
  future generations 8

ice sheets, melting of 49n1
Industrial Revolution 1, 122
infertility scenario 81
intergenerational action, on foreseeable
  risks 59, 62–67, 98, 103, 128; under
  deep uncertainty 73; democratic
  institutions and 132–134; participating
  states engaged in 67; risk-exposed 66;
  risk-imposing 66; on storage of toxic
  nuclear waste 107–109; strong 66;
  weak 66

intergenerational climate justice 2–4, 59, 82, 88; afterlife as a justification of 87; climate change as a matter of 6–10; complexity of 17; duties of 6, 19n5; framework for 16–19; ground rules for 90; human rights approach to 8; interests of future people 87; interpretations of 8; moral challenges posed by 81; norms of 74; obligations of 12, 89; principles of 102; scope of 12–16; theory of 8, 39

intergenerational collective action problem (IGP) 55

intergenerational cooperation: agent's obligations 109; nuanced approach to 111–113; and the scope of justice 109–113; stronger intergenerational actions 111; weaker intergenerational actions 110

intergenerationality/intergenerational: buck-passing 9, 56, 81; climate challenge 10–11, 15–16, 24, 29, 48, 80, 87, 90, 94, 98, 119, 124, 127; connections 65, 74, 126, 132; duties, on climate justice 114, 116, 118, 129; ethics 55, 74; nuances of 113; societies 72, 103; understanding of societies 65

intergenerational justice 80; normative foundations for 82; theory of 82

intergenerational obligations: climate justice as grounds for 68–70, 73; coherence requirement 68, 99; normative requirement 68, 99; on uncertainty and complexity of climate change 80

Intergovernmental Panel on Climate Change (IPCC) 28, 66, 122; Assessment Report 28, 123; on negative impacts of climate change 129

international justice 55

Jamieson, Dale 18, 53–58, 101; on causes and effects of climate change 57; on problem of complexity 73

justice: entitlement 124; judgements of 44, 46; justice-irrelevant risks 47; justice-relevant risks 47; principles of 94; *see also* climate justice

justification for climate justice 91, 94–95
justificatory dependency thesis 96n3

Kantian interpretation of morality 69
Kolodny, Niko 84
Kutz, Christopher 60

livelihood, losses due to climate change 3
long-term climate action plans, implementation of 109
low-technology societies 53

mitigation of climate change 66; burden sharing 86; efforts on a global level 85; future-oriented goal for 67; target of 85

morality: based on respect for agency 69; moral corruption 9, 18, 55, 57, 87, 90; moral response, to climate change 10; moral theories, development of 58, 101

Næss, Arne 54–55
Nationally Determined Contributions (NDCs) 64, 67, 76n8, 107
natural world–unsustainable human behaviour relationship 55
New York, climate change plan 65
non-existence 8–9, 19, 114–119, 127–128
nongovernmental organisations 60
non-identity problem 4–6, 8–9, 13, 19n5, 116–117, 127–128
Notre-Dame de Paris, construction of 63–64
nuclear semiotics 109, 112
nuclear waste depositories 109; *see also* toxic nuclear waste, storage of

obligations of climate justice 9, 25–26, 31, 38, 46, 106, 114, 117, 129, 135; autonomy-relevant conditions 100; changing perspective on 127–134; to future people 73, 89, 124; implications for 40; intergenerational 16, 19n9, 101, 127; to non-existing future people 123; normative requirement 68; organised collective agents 12; to reduce the risk

of climate change 47; scope of 68–69, 110, 114; theoretical issues regarding 127–128
ocean warming 122
O'Neill, Onora 13–16, 60–62, 84, 85, 115, 117
"ongoing enterprise" of humanity 87, 89
organised collective actions, presuppositions behind 61–62
organised collective agents: actions of 69; collective responsibility of 71; duties of 45
organised collectives, relevant to climate justice 59
overinclusiveness 30–32, 47

Page, Edward 58
Parfit, Derek 4, 89
Paris Agreement (2015) 64, 66–67, 70, 86, 103, 107, 112–114, 122, 129, 130; future-oriented mitigation goal 67; intergenerational elements of 67; mitigation objectives of 67; Trump's administration policy to remove the US from 94
pension funds 66
presuppositions, about future people 5, 12–15, 17, 19, 33–38, 40–42, 46–49, 59, 101; about near and remote future generations 99; behind organised collective actions 61–62, 67, 70–73; on capabilities and vulnerabilities 66, 68; on connections between present and future agents 83; definition of 100; intergenerational actions 125; relevant to climate justice 62; reliance on 135
principles of justice 94
public debt, practice of 61, 64–65, 103, 109, 128
public discourse, on climate change 130
public execution 89

quality of life 4–5, 7, 118

reasonableness 43, 45–46
renewable energy: sources of 40; use of 13

requirements of climate justice 13, 41, 61, 118
risk: exposure, within scope of climate justice 37, 42; imposition on climate justice 37, 42, 44; agents imposing 34, 37–38, 41, 44, 66, 100, 125

Scheffler, Samuel 81–82, 84, 89, 91, 132; afterlife conjecture 98, 102–103, 104, 107, 114; argument for future-oriented actions. 93; evaluative dependency thesis 96n3; intergenerational actions 103–105; justificatory dependency thesis 96n3; view of valuing as diachronic 112
scope of climate justice 2, 12–16, 18, 28, 39, 48, 57–58, 80, 118; account of 113–119; afterlife conjecture and 105–106; alternative approach to 35–36; arguments on 124–127; dealing with uncertainty 100–101; foreseeability of 36–38; to future agents 70; for future generations 106–109; intergenerational cooperation and 109–113; non-identity problem and non-existence challenge 114–119; preliminary account of 99–100; taking stock of 98–99
sea levels, increase of 25, 49n1, 66, 129
Shiffrin, Seana 112
Shue, Henry 49n6
social and economic justice 3, 106
social institutions, erosion of 81
society, changing need for justice 115
solutions, to climate challenges: action-centred 59–73; approach for 57–58; coherence requirement 59; collective agents, intentions, and assumptions 59–61; intergenerational actions 62–67; intergenerational obligations 68–70; justice, uncertainty, and complexity 59; normative requirement 59; presuppositions behind organised collective actions 61–62; scrutinising

the presuppositions behind collective
actions 70–73; under uncertainty 59
surface temperature, rise in 122

Thompson, Janna 64
threshold likelihood, condition of
49n6
toxic nuclear waste, storage of
107–110, 113
transgenerational obligations, of climate
justice 64
Trump, Donald 94

uncertainty 53; dealing with 100–101;
future-oriented action under deep
uncertainty 72; situations of deep
uncertainty 71–72
underinclusiveness 47

United Nations Framework Convention on
Climate Change (UNFCCC) 64, 70
use as means 85, 87

Valentini, Laura 7
value/valuing: goods, capability for
112–113; practice of 88, 104, 112–113;
value-laden lives 82
vulnerability of future people, to climate
change 18, 90

warming, safe level of 129
Waste Isolation Pilot Plant in New Mexico,
USA 109
West Antarctic Ice Sheet (WAIS) 28,
31–33, 40–41, 47; melting of 25, 66
White, Lynn 55
Wolf, Clark 8

For Product Safety Concerns and Information please contact our EU
representative GPSR@taylorandfrancis.com
Taylor & Francis Verlag GmbH, Kaufingerstraße 24, 80331 München, Germany

* 9 7 8 1 0 3 2 1 9 3 7 9 3 *